FEATURES

SPRING 2024 • NUMBER 39

T0169579

ESSAY

18

Meeting the Wolf

Greta Gaffin

Should we go back to nature?
Two saints show how to make
peace with it instead.

ESSAY

32

Promised Land

Daniel J. D. Stulac

A newcomer to
Saskatchewan searches
for the biblical promise.

ESSAY

42

**The Language of
the Flowers**

William Thomas Okie

The plants can talk.
Are we listening?

REPORT

54

Saving the Soil

Colin Boller

Can farmers care for
the land and pay the bills?

ESSAY

62

The Wonder of Moths

Caroline Moore

Gorgeous and fragile, moths
showcase nature's richness
and vulnerability.

ESSAY

84

The Leper of Abercuawg

David McBride

In a thousand-year-old
Welsh poem, an outcast
seeks comfort in the wild.

Plough

INSIGHTS

EDITORIAL

10 The Sadness of the Creatures

Should humans live by
the laws of nature?

Peter Mommsen

ESSAY

26 Dandelions: An Apology

The dandelion always
comes back.

Clare Coffey

REFLECTION

40 A Wilderness God

In the Holy Land, the desert
is a place of hope.

Timothy J. Keiderling

PERSONAL HISTORY

50 Why We Hunt

In the woods, to be a
predator is a privilege.

Tim Maendel

READINGS

69 Reading the Book of Nature

Lilias Trotter
Gerard Manley Hopkins
Augustine of Hippo
George MacDonald

DISPATCH

74 Breakwater

In the Channel Islands,
a crumbling jetty protects
a way of life.

Rhys Laverty

PERSONAL HISTORY

92 Lambing Season

I learned some of life's most
important lessons from my
father in the sheep barn.

Norann Voll

DOERS

106 Earthworks

At a community garden
in Detroit, kids help grow
food for the hungry.

Casey Kleczek

ARTS & LETTERS

POETRY

31 Let Them Grow

105 Squall

Robert W. Crawford

READING

80 Are You A Tree?

Or are you a potted plant?

Joy Marie Clarkson

REVIEWS

95 What We're Reading

Andrew Prevot on *Christianity
as a Way of Life*, Lore Ferguson
Wilbert on *Yellowface*, and
Marianne Wright on *Jesus
through Medieval Eyes*.

ESSAY

98 In Defense of Chastity

Is the church's teaching
on sex unnatural?

Erik Varden

DEPARTMENTS

LETTERS

4 **Readers Respond**

FAMILY AND FRIENDS

6 **GEDs for Myanmar Migrants**
Rebecca Newton

9 **Nyansa Classical Community**
Angel Adams Parham

COMMUNITY SNAPSHOT

112 **Watching the Geminids**
Maureen Swinger

FORERUNNERS

118 **Maximus the Confessor**
Susannah Black Roberts

WEB EXCLUSIVES

Read these articles at *plough.com/web39*

ESSAY

Back from Walden Pond
Is Thoreau really the prophet for our
present ecological crisis?
Ian Olson

ESSAY

Ancient Songs in the Desert
For Ephrem the Syrian, nature
shows us truths about its creator.
Tessa Carman

VIEWPOINT

Fixing Men
Masculinity is a good to be praised,
not a problem to be eradicated.
Samuel Helyar

PERSONAL STORY

Nature Hates Me
Spiders, wasps, and mold:
must I love them?
Annemarie Konzelman

Plough

ANOTHER LIFE IS POSSIBLE

EDITOR: Peter Mommsen
SENIOR EDITORS: Shana Goodwin, Maria Hine,
Maureen Swinger, Sam Hine, Susannah Black Roberts
EDITOR-AT-LARGE: Caitrin Keiper
BOOKS AND CULTURE EDITOR: Joy Marie Clarkson
POETRY EDITOR: A. M. Juster
ASSOCIATE EDITORS: Alan Koppschall, Madoc Cairns
CONTRIBUTING EDITORS: Leah Libresco Sargeant,
Brandon McGinley, Jake Meador, Santiago Ramos
UK EDITION: Ian Barth
GERMAN EDITION: Katharina Thonhauser
COPY EDITORS: Wilma Mommsen, Priscilla Jensen
DESIGNERS: Rosalind Stevenson, Miriam Burleson
CREATIVE DIRECTOR: Clare Stober
FACT CHECKER: Suzanne Quinta
MARKETING DIRECTOR: Tim O'Connell
FOUNDING EDITOR: Eberhard Arnold (1883–1935)

Plough Quarterly No. 39: The Riddle of Nature
Published by Plough Publishing House, ISBN 978-1-63608-141-0
Copyright © 2024 by Plough Publishing House. All rights reserved.

EDITORIAL OFFICE
151 Bowne Drive
Walden, NY 12586
T: 845.572.3455
info@plough.com

United Kingdom
Brightling Road
Robertsbridge
TN32 5DR
T: +44(0)1580.883.344

SUBSCRIBER SERVICES
PO Box 8542
Big Sandy, TX 75755
T: 800.521.8011
subscriptions@plough.com

Australia
4188 Gwydir Highway
Elsmore, NSW
2360 Australia
T: +61(0)2.6723.2213

Plough Quarterly (ISSN 2372-2584) is published quarterly by
Plough Publishing House, PO Box 398, Walden, NY 12586.

Individual subscription $36 / £24 / €28 per year.
Subscribers outside of the United States and Canada pay in British pounds or euros.

Periodicals postage paid at Walden, NY 12586 and at additional mailing offices.

POSTMASTER: Send address changes to Plough Quarterly, PO Box 8542, Big Sandy, TX 75755.

Front cover: Albrecht Dürer, *A Young Hare*, watercolor on paper, 1502. Public domain.
Inside front cover: Jean-François Millet, *The Angelus*, oil on canvas, 1857–9. Public domain.
Back cover: Roderick MacIver/Heron Dance Art Studio, *Heron Leaving III Sketch*, watercolor, 2010.
Used by permission.

ABOUT THE COVER

Albrecht Dürer's 1502 painting
A Young Hare is recognized as a
masterpiece in observational art,
but beyond the artistic detail, the
painting challenges the viewer to
ponder the riddle of the bond and
the rift between human and animal.

LETTERS

READERS RESPOND

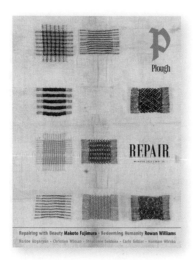

Readers respond to *Plough*'s Winter 2024 issue, *Repair*. Send letters to *letters@plough.com*.

GROUND-LEVEL GRITTINESS

On Christian Wiman's "Ifs Eternally": I like paper and have yet to reach the stage of *enjoyably* reading online. I do so but begrudgingly. I've been reading my first paper copy of *Plough* (Winter 2024). What a pleasure to feel the weight of paper! I first encountered you several years ago through J. Heinrich Arnold's *Discipleship*. I read it, keep a copy close, tell others to buy and read it too, and have given a copy away. Up to now, I've read *Plough* solely online, and I listen to the podcasts but admit to more often than not leaving off before reaching the end. So, finally there's this print copy in my hands. I read it, reread, pick it up yet again. It's quite handsome, too, with fine artwork.

What finally pushed me to subscribe was the prospect of reading an excerpt from Christian Wiman's latest. "Ifs Eternally" wasn't easy and I read it several times, but I got much out of it. Wiman doesn't pretend poetry or living or dying arrives easily. So that drew me in, but then there was Makoto Fujimura and Carlo Gébler and much more. I appreciate the thoughtfully diverse approaches to repair, the often ground-level grittiness, the neighborliness. Who knows, maybe *Plough* can warm three winter months.

Gregory Harris, Clifton, Illinois

OVERCOMING CONSUMERISM

On Peter Mommsen's "In Praise of Repair Culture": Growing up in Evangelicalism, one of the unfortunate lessons I learned was that this life is temporary, which diminishes how we see everything around us. If all we see and know is subject to destruction (or refinement) then what does it matter that our landfills grow exponentially? What does it matter that trash travels around the oceans as floating islands? What does it matter if I destroy my body with sugars and trans fats? Thank you for putting repair and longevity back into the discussion of theology with this article.

Grant Stafford Morgan, Reno, Nevada

I am a part of the repair culture, having a small business in which I repair commercial cleaning equipment. I am appalled at how many perfectly good pieces of equipment have to be destroyed because parts are no longer available for them. I also collect scrap metal as part of my overall business program. It sadly reflects the throwaway society in which we live how many good items I have restored that were given to me as junk. You can blame much of this on Edward Bernays, a father of modern advertising. The whole purpose of ads, Bernays taught, is to make us psychologically dissatisfied with our current condition, and to make us feel that buying the next gaudy trinket will bring us the happiness we seek. And ultimately, who benefits from this consumerist mentality? The same people Jesus chased out of the Temple with a whip – the moneylenders, who make fortunes getting us to buy what we don't need at exorbitant prices.

Edward Hara, Fairfax, Virginia

ZERO EPISCOPALIANS?

On Ben Crosby's "Empty Pews": Well stated! As I finish out my last eight months of ministry (I will be seventy-two in August) in a parish where we have had to learn the meaning of being the church again, I think more and more of our lack of heart for Christ, lack of sharing the gospel, lack of instruction for discipleship. Perhaps it's a tiredness with labels and programs at my age, but nothing galls me more than the annual parochial report, which says little about souls transformed or communities given hope in Christ. It's all about who has the biggest numbers. I pray for you and I pray that the church recommits to its mission and ministry rather than seeking the one magical solution.

Mary Thorpe

Thank you for articulating something I've been clumsily trying to say when explaining why I stopped going to church circa 2018. Until very recently, I wondered if I was

4   *Plough Quarterly*

one of the "dones," but I decided to give it one more shot, with very low expectations, this Advent season. It's hard to be a Christian outside of a community, but so many churches these days make it hard to be a Christian *inside* of a community.

Frances O'Roark Dowell,
Durham, North Carolina

As a Roman Catholic in the American Midwest, I see the same decline in my denomination all around me. I have the same bewilderment with my own denomination's leadership or lack thereof in even merely slowing the decline. Instead of us bemoaning our respective denominations' decline, I ask what fundamental concrete steps this author thinks those in leadership should take? What steps should reasonably active lay people take? Until we can make these steps we're all just engaging in a bunch of navel gazing. I do think in a Western secular society that has a growing epidemic of loneliness and a lack of transcendent beauty, the churches can lean into this and become outposts of beauty and in-person community.

Robert Aitchison, Deerfield, Illinois

REPAIRING WITH CARE

On Kurt Armstrong's "Just Your Handyman": I have not read something this lovely, sincere, and hopeful in a long time. I am so glad that the writer, aside from being a handyman, is also a gifted wordsmith. In a world where bits and bytes dominate, it is so heartening to go through this description of what it takes to be good with your hands. There is so much talk about AI, ChatGPT, and their breed threatening jobs and the livelihoods of millions. I don't think Kurt and his tribe need ever worry on this score.

Arjun Rajagopalan

This really spoke to my heart and soul. As a fellow "doer," I've struggled at times with loving things and accomplishment more than people. My to-do list can easily become an idol. I've felt that my propensity to be task-oriented is at war with my call to love God and my neighbor. But this beautifully crafted article spoke of the harmony that can exist for us doers. The author spoke of integrity, the parts all functioning as a whole. That to me was like a puzzle piece that just fell into place. Be a loving doer however big or small the doing is. "Whatever you do in word or deed do all to the glory of God."

John Geffel, Oregon City, Oregon

THE BENEFITS OF SERVING

On Adrian Pabst's "Why Serve?": I completely agree. A friend of mine is also from Germany and his stint in national service (helping those with addiction issues) led him to a vibrant faith and a lifetime of service to the body of Christ. He was glad that he had a nonmilitary option to pursue. I also agree that it should be compulsory and wonder if there should be military and nonmilitary service options. Want to make America great again? I see all sorts of benefits with this proposal.

Matthew Granitto,
Lakewood, Colorado ➤

About Us

Plough is published by the Bruderhof, an international community of families and singles seeking to follow Jesus together. Members of the Bruderhof are committed to a way of radical discipleship in the spirit of the Sermon on the Mount. Inspired by the first church in Jerusalem (Acts 2 and 4), they renounce private property and share everything in common in a life of nonviolence, justice, and service to neighbors near and far. There are twenty-nine Bruderhof settlements in both rural and urban locations in the United States, England, Germany, Australia, Paraguay, South Korea, and Austria, with around 3000 people in all. To learn more or arrange a visit, see the community's website at *bruderhof.com*.

Plough features original stories, ideas, and culture to inspire faith and action. Starting from the conviction that the teachings and example of Jesus can transform and renew our world, we aim to apply them to all aspects of life, seeking common ground with all people of goodwill regardless of creed. The goal of *Plough* is to build a living network of readers, contributors, and practitioners so that, as we read in Hebrews, we may "spur one another on toward love and good deeds."

Plough includes contributions that we believe are worthy of our readers' consideration, whether or not we fully agree with them. Views expressed by contributors are their own and do not necessarily reflect the editorial position of *Plough* or of the Bruderhof communities. ➤

Photograph courtesy of Partners Relief & Development.

GEDs for Migrants from Myanmar

Partners Relief & Development is working to establish a new educational program in Thailand.

Rebecca Newton

A year ago, I came to Chiang Mai, Thailand, with Partners Relief & Development to help start a GED program for Myanmar migrants. For several years, the organization had provided computer studies and language classes including English, Thai, and Chinese, giving young adults skills that can help them secure better jobs, but the teachers realized that many students had never completed a formal education.

Rebecca Newton is a Bruderhof member working in Thailand.

I spent the first three months recruiting students and developing a curriculum covering the subjects on the American GED exams: math, science, social studies, and language arts. Because the GED exam is recognized by the United States, Canada, Australia, and many Asian countries, it gives the students a chance to apply to university and train in professional fields.

The mission of Partners Relief & Development is to help children affected by conflict live "free, full lives." Projects in Myanmar, Thailand, and the Middle East enable local communities to provide the security and resources children need in order to thrive. That is where the GED program fits in: my students dream of going back to their villages as educators, health care providers, or leaders.

The students' ages range from sixteen to thirty-two; the majority are in their early twenties. Most come from remote farming villages where their parents struggle to raise crops in poor soil and are often unable to sell their produce due to warfare and social instability. At least half of our students left home as schoolchildren, either to live in a town with a school or in a Buddhist monastery as novice monks, because their parents were determined that their children should have a better education.

The SEED School opens at 8:30 a.m., shortly before the students start to arrive by motorbike, bicycle, or foot. My morning class consists of ten students who mostly have jobs during the afternoon or at night. We start by watching a short clip of world news and discussing how it relates to the

Young migrants from Myanmar work toward their GEDs.

subjects we are studying. Then we study until noon.

Our students are used to rote memorization, with corporal punishment meted out for mistakes and questions, so it has been exciting to introduce them to critical thinking, to celebrate questions as the key to learning. Many of the topics we cover, from geography to genetics, are completely new to them, and they are eager to engage with each lesson. The first time a twenty-five-year-old student saw a globe, he told me, "I just want to sit here and look at this all the time." Other students were amazed to learn that our physical appearance is not determined by what we did in a past life, but rather by genetic information passed on from our parents. Examining the founding documents of American government with young people who have grown up under military dictatorship and the chaos of civil war has made me look with new eyes at ideals we in the West take for granted.

I try to incorporate as much group learning as possible, because the students have such a wide range of English abilities and levels of schooling. They often choose to stay on after class and continue studying together for several hours before they have to go to work.

At noon, we take our motorbikes to the local market to buy lunch, and then come back to study some more. On paper, the next few hours are my prep time, but they more often turn into one-on-one tutoring sessions, or a chance to ask individual students how they are doing and try to encourage them through the various challenges they face, either academically or out of school. Toward the end of the afternoon, I take a break, because three times a week I come back to teach an evening group of GED students, who mostly work six if not seven days a week in construction, restaurants, gas stations, or factories.

Even if I am feeling tired, the minute the students walk in, their enthusiasm reenergizes me. They look out for each other and tease each other like siblings. Because their work schedules keep them so busy, school is both education and social life.

It is rewarding to teach students who are so hungry to learn. I have gained insights into their culture and the incredible obstacles they have overcome: losing parents to sicknesses that should have been treatable, seeing fathers turn to alcoholism when they lost everything in military raids, leaving home as six-year-olds because their parents were determined that they should get an education, moving schools multiple times as the fronts in the civil war shifted, uprooting from their home and culture and trying to make their way as second-class citizens in a country with a different language and customs, working long hours to support not only themselves but their families back in Myanmar.

These young people have stepped into such heavy responsibilities that they have missed out on much of the fun and friendships and opportunities that I associate with my teen years. I try to weave those opportunities into our program with hands-on experiments and projects, occasional class trips, or a chance to cook and share a meal together.

One of my new students, a sixteen-year-old, had missed some classes, having difficulty traveling to Thailand from Myanmar, so I took a few hours one afternoon to work through some math and social studies concepts. At the end, I asked him, "Do you have any more questions?" He was quiet, and I could see him searching for words to formulate his thought. Then he asked me, "How can we teach our children to grow up to be better people?"

Caught off guard, I acknowledged that this was a big question, but that to my mind, the most important way forward is to lead by example, to model compassion, and to teach others to think for themselves (something that is discouraged under his homeland's dictatorship). I asked, "Do you want to be a teacher?" He responded, "I want to be a teacher and a leader, because there are so many things that are not good in our country, and I want to help my people."

These are the visionaries I am privileged to walk alongside as they study and work and hope for a brighter future for their people.

Rebecca Newton, *left*, with some of her students.

Photographs courtesy of Partners Relief & Development

My name is Sai Saw. I was born in a small village in Shan State. We are six people in my family. My parents are farmers. We lived in a small bamboo house near the mountain. There were no hospitals or schools and no electricity in my village. When I was young, I didn't have the opportunity to attend school, because it was very far from my village and if you want to attend, you have to pay a lot of money. One day in the morning when I was feeding buffaloes on the farmland, just like every day, my mother said to me, if you want to learn how to read and write in Shan, I will take you to live at the temple. Five days later she took me to the temple near our village. I lived there almost ten years.

One morning the Burmese soldiers came to my village. They killed our buffalo and our cows, and burned down our house. The situation was very bad, so my mother sent me to live with my brother in Thailand.

My first job in Thailand was at a gas station. After two years, I became a factory worker. I worked in the daytime, and in the evening, I came to study English at SEED School, but it was very difficult for me, because in the past I had never been to school. I think I am a lucky person that I found SEED School.

I wanted to join the GED program, because I want to improve my English. I also want to meet new people and learn new ideas. After I finish my education, I want to bring the knowledge I learn from the university and go back to teach the children in my village. I hope I can do that.

My name is Nan Hla. I am seventeen years old and I was born in Tachileik. When I was five years old I moved to Hekel Village in Myanmar. I studied at Hekel School until seventh grade. In 2018 I came to Thailand because my mother and grandmother had problems, but I didn't want to come to Thailand because I knew I would have to work. I wanted to continue studying, so my mother said she would let me study in Thailand and she brought me.

When I arrived, I didn't actually study, because my family didn't have the money to send me to school. That is why I have to work. In 2022, my friend recommended SEED School to me. Then I applied to study English and Chinese. I work during the day and go to school in the evening. I started studying English and now I am studying GED in the evenings because I want to go to university. I think it is suitable for me because I can study while working. I can also follow my dream.

My name is Sai Seng Li. I am sixteen. I grew up in a small village far from the town: Namlan, Northern Shan State, Myanmar. There are seven people in my family. There are my grandparents, my parents, and two siblings. My parents are farmers. And this is about my journey of education. I began my education in my village when I was five years old. Later, I moved to live in a town to attend the government school at the age of nine, spending five years there, until grade seven. At that time Covid had come, so the school did not open.

I made up my mind to attend the national school at age fourteen, but I had to repeat grade seven. However, my teacher, my friends and I could not live in a single place for a long time. Due to the conflict between different groups of soldiers, we often had to move to a new place where it is peaceful and safe. I attended that school for three years in four different places.

I have a dream to become a qualified teacher. I realized there are not enough teachers in my country. Moreover, there are so many reasons why I want to be a teacher. As a result, now I am in Chiang Mai studying at SEED School. Attending the GED will help me achieve my specific goal and help me improve skills for my future. If I did not meet those teachers who helped me, I would not have this day. Finally, after I finish the GED I believe I could be a small brightness for my environment.

For more about Partners Relief & Development, visit www.partners.ngo. ➤

Inside Nyansa Classical Community

Nyansa cultivates knowledge and wisdom, merging classics and culture to reach the next generation.

Angel Adams Parham

Nyansa Classical Community was founded to bring classical learning and literature to young people of diverse backgrounds who are unlikely to be introduced to this tradition in any other way. Because much of our work has been with children of African descent, it was important to us early on to weave together the classics with the history and culture of Black authors and artists.

At a gala we held to introduce friends and donors to our work, my college-student assistants had assembled examples of our children's artwork compiled as we read Homer's *Odyssey*. The students had the chance to combine two streams of art: the classical tradition of Homer and his great epics and Black intellectual and artistic traditions. They created collage art in the style of Romare Bearden's *A Black Odyssey* – a beautiful series which recasts Homer's *Odyssey* by telling the story of African and African-diaspora people in our various journeys across the ocean and in the new places we call home. Bearden, an African-American painter of the Harlem Renaissance, brings a beauty and mastery to this work that is a treasure to pass down to new generations.

After operating as an afterschool program for six years, the Nyansa program came to

---

Angel Adams Parham is the executive director of Nyansa Classical Community.

a halt with the Covid pandemic. We used this time to distill all that we had been teaching into a formalized curriculum divided into twenty weeks. This would make it possible for others to replicate the best of what we had been practicing. Writers and artists drew together to describe on the page what we had been experiencing in person.

By the fall of 2021, we were in a position to pilot the curriculum at several different sites. Over the course of the academic year it was used in two schools in Virginia and one in Uganda. One of the most gratifying reports came from Sarah, a teacher who, on her own time after school, took on the challenge of working with two brothers, fifteen and twelve, who were in academic trouble.

Even though the material was pitched at an elementary-school audience, Sarah was able to use it as an educational breakthrough with the fifteen-year-old, who was behind in reading and exhibited little interest in school. She drew on Nyansa's story-rich curriculum – in addition to Greek mythology, there are Bible stories and Aesop fables – to engage him. Following each story were writing assignments based on the literature that helped to reinforce areas of reading, spelling, and writing. The younger brother, for his part, was most attracted to the art project based on African-American artist Jacob

Lawrence's *The Migration Series*. All of these lessons integrate literature, history, and hands-on art. The stories behind Lawrence's artwork allowed Sarah to draw this young student into conversations about virtue and vice and consider how to cultivate the one while fleeing from the other.

Nyansa can travel to where young people are, whether in a living room after school, working with a tutor, or at a school in urban Virginia or rural Uganda. We are grateful to be part of passing on this great tradition and helping young people find their voices so they can join this millennia-long conversation.

From The Liberating Arts *(Plough 2023).* nyansaclassicalcommunity.org ➤

Poet in This Issue:

Robert W. Crawford has published two books of poetry, *The Empty Chair* (2011), and *Too Much Explanation Can Ruin a Man* (2005). His sonnets have twice won the Howard Nemerov Sonnet Award. He lives in Chester, New Hampshire, and is the Director of Frost Farm Poetry at the Robert Frost Farm in Derry, New Hampshire, and is responsible for the Hyla Brook Reading Series. Read his poems "Let Them Grow" on page 31 and "Squall" on page 105. ➤

The Sadness of the Creatures

Should humans live by the laws of nature?

PETER MOMMSEN

THE DAY I BUTCHERED my first lamb was one of those hot but brilliant September Sundays. My friend Achim had come from the next village over – our family was then living in rural Thuringia – to supervise and coach. A former East German police officer, Achim had his own flock and decades of experience, as well as the required bolt gun. We were living on the meager income from a start-up landscaping business at the time, and there was never enough protein in the house. But there were a couple of unused pastures next to us, so it made sense to get some sheep and goats. After a string of newbie-farmer setbacks – some of the animals died after being fed leftover birthday

Albrecht Dürer, *Head of a Stag*, watercolor, ca. 1503.

cake by a well-meaning neighbor – it was finally time to get some meat.

Achim believed in the old ways and had brought his traditional wooden slaughtering bench, setting it up on the pavement by the garage over a storm drain. He helped bring up the first victim. At seven months, the animal had lost the loveliness of early lambhood for the strongly scented boorishness of an almost-sheep. But as we heaved him up onto the plank, before the bolt gun started the process that would turn him into crown rib and shoulder roast, I glimpsed his eyes, moist and beautiful.

That evening we feasted together on Achim's specialty – stomach flap of lamb rolled up like a crepe with the sliced kidneys inside, seasoned with loads of caraway before braising in beer – plus his homemade cherry wine. It was a good day. But the look in the animal's eyes stayed with me, and with it the sense of connection to a fellow living creature whose life was sacrificed for my benefit. The poet W. H. Auden captured this feeling hauntingly:

> Our hunting fathers told the story
>   Of the sadness of the creatures,
> Pitied the limits and the lack
>   Set in their finished features;
> Saw in the lion's intolerant look,
> Behind the quarry's dying glare,
> Love raging for the personal glory
>   That reason's gift would add . . .

A sense of connection in our shared experience of conscious life. But also an unbridgeable distance between the animal's world and mine.

Today, the sacrifices that humans extract from sad creatures are largely out of sight and mind, and with this remove, our connection to them. Our daily lives have become ever more urbanized and screen-bound; only 4 percent of Europeans and 2 percent of Americans are full-time farmers.

Yet where connection grows thin, the significance of what sets us apart from the rest of nature slips from sight as well. Many moderns are unsure of what difference, if any, marks us out from other living beings on our planet, and of what our place in the natural world ought to be.

What can nature itself tell us about how to live within it?

SINCE THE INDUSTRIAL REVOLUTION, our species, flush with technological mastery, has exercised unprecedented dominance over nature, with consequences that are now catching up with us. The litany of ills is by now a cliché, though no less true for that: pollution, deforestation, habitat loss, climate change, a mass extinction unequaled since the Cretaceous. Add to that list the effects of the industrial agriculture that feeds us, including soil exhaustion, poisoned aquifers, and "concentrated animal feeding operations" in which hundreds of millions of animals are held, a system far removed from Achim's style of farming. Even our dogs seem to suffer from high rates of anxiety, according to a 2019 study published in the *Journal of Veterinary Behavior*, likely in part because of their unnaturally confined and sedentary lifestyle.

Over the past decades, an environmentalist movement has risen in protest against these destructive habits. Less clear than the fact of exploitation is the solution. Some advocate harnessing technology and finance to promote sustainability. Others urge us to abandon technological capitalism and pursue degrowth. Not a few argue that there are simply too many humans – that population decline, or even human extinction, would be good news for the planet.

Since the industrialized West bears chief responsibility for environmental collapse, many point to the West's historically dominant religion, Christianity, as a key culprit. This helps explain why paganism is one of the fastest-growing faiths in erstwhile Christendom, with strong appeal to Millennials and Gen Z; according to a 2014 Pew study, around one million Americans identify as pagans or witches, up from just a few thousand in 1990. In its critics' view, Christianity is guilty of using the Genesis command to "fill the earth and subdue it" as a license to exploit. Its alleged dualism, privileging the soul over the flesh, has led to contempt for the body and for biological life more generally, claiming human exceptionalism where it should see the unity of life.

Paganism offers an alluring alternative. We humans, pagans suggest, are not to claim arrogant superiority over nature – we're just part of it. As a self-identified Green Witch explained to *Quartz*, her faith involves "a deep adherence to nature and natural law, an attention to the cycles of the earth and the lives within it." Nature is charged with divine power, as ancient pantheism taught; whatever god there is dwells within it, not outside it. Talk of human exceptionalism obscures our kinship with other living beings.

These ideas, which date back to the pre-Christian era, resurfaced in the nineteenth century and have grown in parallel to industrialization. In 1939, for example, such ideas were advanced by an anonymous editorialist:

> To us, God is manifest everywhere in nature, because nature is sacred, and we worship in it the revelation of an eternal will. Seen in this light, the animal is, in our eyes, actually a "little brother," and our sensibility considers that assaulting a man able to defend himself is more morally acceptable than any cruelty towards a defenseless creature.

These reflections appeared in the SS journal *Schwarze Korps*, as quoted in Johann Chapoutot's 2018 *The Law of the Blood: Thinking and Acting as a Nazi*. The book is a fascinating exploration of how such a tender idea – the sacredness of nature, animals as our little brothers – was used to justify the least tender and sacred behavior ever known.

Not that the Nazis' conclusions are inevitable, as we can see from the variety of modern paganisms, most of them simply seeking oneness with nature's harmony. But nature is also harsh and brutal, and an ideology that "worship[s] in it the revelation of an eternal will" opens itself to embracing the dark side of its law.

The Nazis focused on certain scientific facts that a green paganism would prefer not to see. The main lesson they drew from nature was one of systematic cruelty: the domination of the weak by the strong, the elimination of the unfit, the merciless competition for survival. "All Life Is Struggle" (*Alles Leben ist Kampf*) is the title of a 1937 propaganda film that promotes eugenics and a sterilization campaign. The film aims to rid viewers of any residual Christian conditioning that might tempt them to protect the vulnerable. Between scenes of fighting stags, monkeys, and boars, intertitles admonish: "Only the *best* genes are passed on. . . . What is weak or unfit for life must succumb to the strong. Nature allows only the *best* life forces to survive."

It's a grim vision. Yet when judged by the evidence of evolutionary biology, the film's understanding of nature gets nearer to reality than does that of modern solstice celebrants at Stonehenge. Fatefully, it makes the additional step of counseling its audience to live in accordance with nature quite literally – to adopt its ruthless ways as their own. Contrary to the teaching of the Abrahamic religions, it suggests that humans are not exceptions to the law of survival of the fittest, nor should they want to be. The Christian doctrine of dominion, in which humankind's unique calling is to act as creation's steward in God's stead, is rejected in favor of embracing our biological drives. Heinrich Himmler expressed these ideas in 1942:

> It is time to break with the folly of these megalomaniacs, in particular these Christians, who speak of dominating the earth; all of that must be brought back into perspective. There is nothing particular about man. He is but a part of this world. . . . Man must relearn how to see the world with worshipful respect.

This would sound positively humble if one forgets that Himmler was then commanding *Einsatztruppen* on the East Front in a genocidal bid to expand the German people's habitat (*Lebensraum*). His "worshipful respect" involved shedding "anti-nature" inhibitions such as pity for the victim, grounded as they are in supernatural beliefs. As one likeminded professor of eugenics put it in a 1937 lecture: "We are all a part of nature, we result from nature's law. Why should our intelligence deviate from understanding nature's laws to explore any kind of 'metaphysics,' anything 'supernatural'?" Nature is all there is, and from its law there is no appeal.

Similar writers traced Christianity's belief in the supernatural and its supposed contempt of nature to its Jewish inheritance. As the SS leader Richard Walther Darré explained, Jews and Christians share a faith in "Yahweh, the vengeful, Eastern, non-native god of the deserts, come to devastate the forests and lakes of verdant Europe." Like an invasive microbe, the body-denying spirituality of Jewish-derived Christianity threatened to destroy indigenous Europeans' joy in bodily life.

It's worth asking if there are any lessons to draw. Here, after all, is the most vivid test case of a formerly Christian society that chose to revert to nature's law. It illustrates the outer implications of a certain form of paganism. If one denies any distance between humankind and nature, it's hard to see what could be wrong with "might makes right."

In an increasingly post-Christian age, this history has renewed salience. To be sure, the 2020s are not the 1930s, and to suggest tidy equivalences would be foolish. But certain strains run through. The belief that unfit lives are not worth living, for example, has regained a foothold through the prenatal elimination of babies with Down syndrome and the euthanasia of people with disabilities; both practices are now widespread on

both sides of the Atlantic, typically championed by progressives with environmentalist sympathies. For its part, the new right with its Bronze Age fantasies of "sun and steel" has few objections to the new eugenics either. Its exponents champion the strong over the weak, mock Christianity's solicitude for the vulnerable, and obsess over supposed racial distinctives, including by reviving old-fashioned anti-Semitism. The de-Christianized "law of nature" has a knack for reappearing in new times and guises.

CONTRA ITS CRITICS, not to mention its careless practitioners, Christianity's true relationship to nature is not one of contempt or disassociation. According to one ancient church tradition, while nature must not be our source as a law to live by, it still bears a different kind of meaning: as a book to be read. This image goes

Albrecht Dürer, *The Little Owl*, watercolor, 1506.

laid aside for too long, quickly ceases to hang together and make sense. The baffled reader is forced to start again from the beginning.

That is my experience, anyway. In my mid-twenties, after three years spent mostly indoors with other people or my laptop (I'd been running a magazine), I suddenly found myself with hours of free time alone. After work, I'd go daily into the woods of southwestern Pennsylvania. Most days, somewhere along the trail I'd run into Arthur Woolston, an English naturalist and fellow Bruderhof member then in his eighties.

Arthur was a slight, stooped man with a short white beard, a pair of binoculars, and a face filled with delight. We'd stop for five minutes or half an hour, and he'd point out something about the woods around us that I'd either forgotten or never known: the species of that fern or fungus, the identity of an unseen hermit thrush, how to tell the difference between a Norway and Sugar maple. His family, I knew, worried about his long solitary rambles – he had a heart condition – but no one dared ask him to stop. The intensity of his joy in the natural world seemed concentrated by his awareness that he would soon have to leave it.

Arthur was a walking Peterson field guide to birds, trees, insects, and animal tracks. His relatives once lent me his unpublished autobiography, four hundred typed pages describing his itinerant life in Britain and South and North America. His account tended to pass over the milestones that most people construct their life stories around. Instead, it would dwell lovingly on a dozen birds

back at least as far as a third-century Desert Father, Anthony the Great.[1] Anthony was an illiterate former swineherd who spent fifteen years as a hermit in the Egyptian wilderness before founding one of the first Christian monasteries. Because he could not read scripture, Anthony turned to nature: "My book is the created nature, one always at my disposal whenever I want to read God's words."

If we moderns have trouble reading the book of nature, perhaps that's because, unlike our pre-industrial forebears, we lack practice. Like a Tolstoyan novel with dozens of characters and an intricately interwoven plot, nature's book, if

1. I'm drawing on the research of the Italian astronomer Giuseppe Tanzella-Nitti, who traces the metaphor's development in "The Two Books Prior to the Scientific Revolution" (2005) and "Jesus Christ, Incarnation and Doctrine of Logos" (2008).

Albrecht Dürer, *The Great Piece of Turf*, watercolor, pen and ink, 1503.

and wildflowers he'd seen one day in 1946 while walking through the Paraguayan jungle to his job at a sawmill, giving the species in both English and Latin. Learning to name the creatures, Arthur knew, can be the first step to paying them proper attention. The reward is the thrill of recognition, even of kinship.

TO READ THE BOOK OF NATURE, you have to actually pay attention – go out and fill your gaze with the stars, or a forest, or a deer. Without this practice, the book is illegible, and it is difficult to make sense of the Psalmist's claim that the "the heavens declare the glory of God, and the sky above proclaims his handiwork." Similarly, it will be hard to follow the apostle Paul's argument that "what can be known about God is plain to [human beings], because God has shown it to them. For his invisible attributes, namely, his eternal power and divine nature, have been clearly perceived, ever since the creation of the world, in the things that have been made."

How can that be if nature is as cruel as evolutionary biology tells us? Far from manifesting peaceful unity, it is as "red in tooth and claw" as Darwin described, possessing not only beauty and order but also pathogens and parasites. "All life is struggle" is only partly false; the ruthlessness the Nazis admired is undeniably there.

Though he predated Darwin by fourteen centuries, Augustine of Hippo wrestled with similar questions, wondering in his *Confessions*, for example, how a good God could have created repulsive insects. It's worth noting that Augustine never regarded the book of nature as showing a static perfection, all harmony and innocence. Instead – to paraphrase Rowan Williams – Augustine described natural phenomena as emerging from a world in flux, where nothing other than God is changeless or deathless, and competing forces are always at work upon one another. We must acknowledge, Augustine thought, that not everything in nature is orderly, purposive, or beautiful; yet nature shows a remarkable *tendency* toward order, purpose, and beauty, as if drawn toward them. It is in this tendency, he believed, that we perceive the hand of a good Creator.

In science, the intelligibility of nature, its tendency toward order, is itself a wonder. "It could be said that the eternal mystery of the world is its

---

If we moderns have trouble reading the book of nature, perhaps that's because, unlike our pre-industrial forebears, we lack practice.

---

comprehensibility," said Einstein. "The fact that it is comprehensible is truly a miracle." There's the remarkable fact, for example, that the universe is not arbitrary, as it conceivably might be, but runs according to discernible natural laws that are (as far as we know) valid everywhere and always. For instance, the maximum speed of light, the laws of gravity, and the mass of an electron are apparently the same anywhere in the universe and at any stage of its development. These laws, further, can be described by mathematics – a logically coherent and purely intellectual system that nevertheless corresponds to actual reality. According to the physicist Eugene Wigner, "The enormous usefulness of mathematics in the natural sciences is something bordering on the mysterious. . . . There is no rational explanation for it."

Such considerations, of course, don't prove the Christian view of nature to be true. But they are consonant with it. One of Christianity's core claims is that the universe was created by the Logos. The Greek term – often translated as "word," though it has many meanings including "reason" – derives from philosophers going back through Plato to Heraclitus. Already early on, they regarded the Logos as divine. As a creative principle of intelligibility and harmony, it gave shape to the cosmos. At the time the New Testament was written, notes the scholar Giuseppe Tanzella-Nitti, in Greek philosophy the terms "Logos, Artificer of creation, and Soul of the world [had] all become synonyms referring to God."

This is the meaning of "the Word" when it appears at the beginning of the Gospel of John: "In the beginning was the Logos, and the Logos was with God, and the Logos was God. He was

---

The Logos is not only the cosmic creator of galaxies, muons, and general relativity; he is also a man of sorrows who suffered and died, and whose sign is the paschal lamb.

---

in the beginning with God. All things were made through him, and without him was not anything made that was made." Nature is the creation of a Reason who preceded it and transcends it.

So far John is expanding on the Greeks. But then he goes where they never went: "And the Logos became flesh and dwelt among us, and we have seen his glory, glory as of the only Son from the Father, full of grace and truth." The Logos has entered personally into his creation as the flesh-and-blood human being Jesus of Nazareth, while still remaining who he is. That is, he crossed the distance between God and man, one seemingly greater than that between man and animal. When we open the book of nature, this is the Word we are to read.

THAT INSIGHT WAS CENTRAL to the thinking of the Radical Reformer Hans Hut, whose signature theme was "the gospel of all the creatures." To him, the problem of suffering does not complicate the book of nature but rather is the key to understanding it.

Hut was a traveling bookseller from Thuringia who fought as a partisan in the German Peasant Revolt of 1525. Surviving the massacre of peasants that ended the rebellion, he joined the Anabaptist movement soon afterward, in his late thirties. He was soon traveling widely across central Europe as an Anabaptist missionary at a time when capture meant likely death. Arrested in 1527 and subjected to torture, he died after three months in a prison accident; the next day his corpse was dragged to court, formally sentenced to death, and consigned to the pyre.

Hut was a mystic in the medieval tradition of Thomas à Kempis and Meister Eckhart, who taught the imitation of Christ in daily life. His brief underground ministry of sixteen months was extraordinarily productive, including authorship of letters and poems from which we can glean the content of his preaching.

Like Anthony the Great, Hut emphasized that the gospel of all the creatures, unlike the written Bible, was accessible to everyone, literate and illiterate, rich and poor. The peculiar power of Hut's approach is his emphasis on the passion. The creatures preach to us not primarily through their order, purpose, or beauty, but through their suffering: "Noteworthy in all [Christ's] parables is that the creatures are made to suffer the effects of human activity. It is through this pain that they reach their goal, that is, what they were created for. . . . If one wants to use an animal, it must first be dealt with according to human will; it must be prepared, cooked, and roasted. That is, the animal must suffer."

The passion of the creatures must have had special resonance for Hut, who knew that his comrades, including one of his daughters, were being executed one by one, and who couldn't expect to live long himself. The lamb of sacrifice is never far from his mind. Of course, Hut

acknowledged, the animals suffer involuntarily, while his own readiness for self-sacrifice was chosen. Still, to him there was an analogy. In animals' suffering, he saw a symbol of what the imitation of Christ demands: "In the gospel of the creatures, nothing is signified and preached other than Christ the crucified alone. . . . Preaching this Christ is what all the creatures teach."

Hut's gospel points to a deeper way to read the book of nature. The Logos is not only the cosmic creator of galaxies, muons, and general relativity; he is also a man of sorrows who suffered and died, and whose sign is the paschal lamb. In him, the sadness of the creatures is not annulled – at least, not yet. In John's Apocalypse, a climactic scene involves "every creature in heaven and on earth and under the earth and in the sea, and all that is in them." They gather in a vast multitude around the divine throne, next to which stands the Logos, now a Lamb. He is given a scroll – a book with answers to the riddle of creation's suffering, which only he can open. In one voice, the creatures announce that glory and power belong to the Lamb who was slain. ➤

Albrecht Dürer, *Wing of a Blue Roller*, watercolor on vellum, 1512.

Meeting the Wolf

Should we go back to nature? Two saints show how to make peace with it instead.

GRETA GAFFIN

THE LAST WOLF IN BOSTON was killed in 1657. I live in Boston, a world of stone and brick and metal and glass. Many modern people live in similar surroundings.

And we desire to escape. We want to flee this manmade environment – loud and overstimulating and chaotic – and retreat to what is peaceful and calm.

Some of our overstimulation comes from a barrage of messages about nature. We're bombarded with them. Natural living. Returning to the land. "It's just so important to me to get out into nature" – you hear it from both secular and religious people; it's put in terms of mental or spiritual health or both.

I live in this society, and I want to do this too. But what does it mean to "get into nature"?

Recently I went on a retreat at the Society of Saint John the Evangelist, an Episcopal monastery in Cambridge, Massachusetts. I took some time to walk along the street in front of the monastery, enjoying the plants. Nature, after all. The thought came to me that it is very nice of the Cambridge Department of Public Works to maintain the plants for me to enjoy.

The pleasant rolling hill near where I grew up has that shape because the hill is built on top of what was once Boston's landfill. The city forest has paths maintained by the Department of Conservation and Recreation. Untrammeled nature? Maybe not. Beautiful? Healing? Absolutely.

But that's Boston. I was also, briefly, an intern at the Community of Saint Mary, another Episcopal religious house in Sewanee, Tennessee, a long way away from urban life. It got a lot darker there than I was used to. I stayed in a little cottage down the path from the convent, and the gardener warned me that there might be snakes. I was afraid walking back from the main convent building to the cottage: What if I stepped on one? Snakes in rural Tennessee are not like Boston's humble garter snake. If you drove out from the convent into the nearby hollows, you would find Christians who take up serpents because the Bible tells them to, and sometimes they die.

A couple of years later I spent a month in rural Ireland and found myself again walking down poorly lit rural paths at night, once again worrying about snakes. But each time I would stop myself and remember: Saint Patrick got rid of those. Not like in Tennessee. Natural historians tell us that there had in fact been no snakes in Ireland since well before Patrick arrived. Natural historians would know. But I thought of Patrick nonetheless. For someone who arrived in Ireland from Great Britain or Scandinavia, where there are poisonous snakes, not having any really would've been miraculous.

SAINT FRANCIS IS NOW REMEMBERED as a charming patron saint of animals, a man who preached to the birds. People bring their pets to church once a year on his feast day to be blessed. On Saint Francis's feast day, even the Episcopal bishop of East Tennessee might handle serpents . . . to bless them as pets.

In his own time, though, Francis was beloved for his taming of *wild* animals. The story goes that around 1220, a wolf was terrorizing the Italian city of Gubbio. It snapped up livestock and eventually graduated to humans. They became his preferred meals, and at last, no human who dared venture outside the city walls was safe. Francis, who was living in the city at the time, told the townsfolk he was going to parley with the wolf, and passed beyond the city walls, citizens trailing a little way behind. He went up to the wolf's lair. When it came out, it made as if to devour him, but he made the sign of the cross and commanded it in the name of God to stop its depredations. It meekly laid its head at his feet. But Francis was not quite done. "Brother wolf," he said – the citizens were close enough to hear this part –

> thou hast done much evil in this land, destroying and killing the creatures of God without his permission; yea, not animals only hast thou destroyed, but thou hast even dared to devour men, made after the image of God; for which thing thou art worthy of being hanged like a robber and a murderer. All men cry out against thee, the dogs pursue thee, and all the inhabitants of this city are thy enemies; but I will make peace between them and thee, O brother wolf, if so be thou no

Greta Gaffin is a freelance writer from Boston, Massachusetts. She has a master of theological studies degree from Boston University and a bachelor's in economics from the University of Massachusetts Amherst.

Previous spread: European wolf. All photography by Danny Green.

more offend them, and they shall forgive thee all thy past offenses, and neither men nor dogs shall pursue thee any more.

The wolf indicated his agreement to these terms, bowing his head again to Francis. "As thou art willing to make this peace," said the saint,

> I promise thee that thou shalt be fed every day by the inhabitants of this land so long as thou shalt live among them; thou shalt no longer suffer hunger, as it is hunger which has made thee do so much evil; but if I obtain all this for thee, thou must promise, on thy side, never again to attack any animal or any human being; dost thou make this promise?

Francis stretched out his hand, and the wolf placed his forepaw in it, sealing the oath between them.

Today we care deeply about conservation. We care about the conservation of wolves. In medieval Europe, and in North America until the twentieth century, people tended to care more about the conservation of their livestock, and of their own skins. Wolves were dangerous to both.

This is why the last wolf in Boston was killed in 1657. But one night last year, I was walking from the Boston University School of Theology to my dorm room on BU's Fenway campus. That involves walking past part of the Emerald Necklace, the chain of local parks designed by Frederick Law Olmsted, who also designed Central Park. Everything there is meant to be there, just as Olmsted planned it, including the parts that are meant to look wild.

And there was a coyote.

Coyotes do not eat adult humans. I was not in danger. But I was afraid.

And I was in part afraid because I was not in "nature." I was two blocks from Fenway Park and new, gleaming skyscrapers and the sleek lab buildings of BU's attempt to become the biotech

Wolf in autumn.

Wolf and reflection, No Man's Land, Finland–Russia border.

hub of the United States. You *go out into* nature; you choose to be there. And I had not chosen to be there. I had chosen to be here.

The coyote had presumably chosen to be here as well, with or without my permission, or the state's. He was presumably here to hunt rabbits, the adorable bunnies I enjoyed seeing hop all over the campus's neatly manicured lawns. But you don't have rabbits without predators – not naturally, anyway.

In Leviticus 26, God says to the Israelites: "If you follow my statutes and keep my commandments and observe them faithfully . . . I will grant peace in the land, and you shall lie down, and no one shall make you afraid; I will remove dangerous animals from the land, and no sword shall go through your land." Removing wild beasts, then, is regarded as in the same category as ending war.

The most common wild animal an urban dweller sees today is, if not the rat, the pigeon. But they're feral (that is, descended from the domesticated bird), so they're not really wild. Or it might be the squirrel. Olmsted and his ilk intentionally reintroduced the squirrel to urban parks in the nineteenth century. Architects like Olmsted saw the dirty, industrialized world a growing percentage of Americans lived in, and wanted to reintroduce beauty and a simple, rustic character. The squirrel is harmless. It may be annoying when it digs up your garden, but if you live in Boston, your garden, if you have one, is not your sole source of food. You may as an urban gardener feel that you are at war with the squirrels. But you're really not. Not the kind of war that God would have to step in and stop.

THE DESERT FATHERS went out into the desert because it was wild and because *nobody else wanted to be there.* It was not like the pleasant retreat we might take in the Adirondacks. It was very much a sacrifice to move there.

The problem was not just the wild beasts of the desert, although they were a problem too, but the landscape itself. The people of Egypt have huddled along the banks of the Nile for thousands of years for a reason: its water brings survival. The farther one gets from the river, the harder it is to survive. To live in the desert, to live "in nature" and outside of civilization, is to be on the precipice of death. Jesus went there. The Israelites passed through it.

And yet this way of life was appealing, perhaps especially to those who could've easily chosen not to live it. Saint Anthony of the Desert, one of the early monastics, grew up in a wealthy family. After his parents died, he heeded Matthew 19:21: "If you want to be perfect, go, sell what you have and give to the poor, and you will have treasures in heaven." That is what he did. He sold his family's property. He went into the desert. He did not want acolytes. He had them anyway. They saw his holiness, which followed from his going out into nature, because that going out into nature was a meaningful sacrifice for Christ.

One of his modern acolytes – in a way – is Chris McCandless, immortalized in Jon Krakauer's bestseller *Into the Wild.* After graduating from Emory University, McCandless gave the rest of his college fund to charity. He became a wanderer. He eventually decided to move to the Alaskan bush with no supplies, hoping to live off the land. He found an abandoned school bus to live in. He lived there for 113 days. Then he died of starvation.

Like the graves of many early saints, the bus became a shrine. Pilgrims visited from all over the world hoping to retrace his steps, to the point where the state of Alaska removed the bus. There had been too many search-and-rescue missions to save visitors who were themselves about to become victims of nature.

The writer Bobby Angel describes McCandless as "a Saint Francis who had never encountered Christ." There's that difference between McCandless and Saint Anthony as well, but there's another one also. Anthony did not live a totally isolated life. He did not sustain himself off the land. If he had tried, he would have died, just as McCandless did.

McCandless was inspired not by those early saints but by Henry David Thoreau, who "went to the woods" because he

> wished to live deliberately, to front only the essential facts of life, and see if I could not learn what it had to teach, and not, when I came to die, discover that I had not lived. I did not wish to live what was not life, living is so dear; nor did I wish to practice resignation, unless it was quite necessary. I wanted to live deep and suck out all the marrow of life, to live so sturdily and Spartan-like as to put to rout all that was not life, to cut a broad swath and shave close, to drive life into a corner, and reduce it to its lowest terms.

But Thoreau, of course, was writing from Walden Pond, from the hut he built on his friend Ralph Waldo Emerson's land. It was around a twenty-minute walk to his mother's house. Over the two years that he lived there, she did his laundry. The train into Boston ran along one side of the pond. Thoreau spent his time writing, as well as doing some land-clearing for Emerson in lieu of rent. Anthony spent a significant amount of his time weaving baskets to exchange for food and other necessities. Other desert monks did too, and many of them complained about having to spend so much time weaving compared to prayer.

The Sayings of the Desert Fathers, a selection of stories about the earliest hermits, gives an anecdote about Abba John the Dwarf. He told his older brother he wanted to be like an angel, who did not have to work but could ceaselessly praise God. He went to the desert. Within a week he was back, and his brother told him that he was a man, not an angel, and men have to work to eat.

One of the appeals of "going into nature" is getting away from the rat race. But it's hard to sustain yourself out there – a lot harder than sending emails. Anthony would have had a much easier life if he had kept his family's land and overseen its cultivation. He gave up a lot to live this life of prayer in the desert. So did Francis, who famously tore off his expensive aristocrat's clothing in the streets of Assisi, renouncing his patrimony along with his tunic, leaving them to

---

Our ideas about nature are often drawn from Romanticism, from the writings of men of means who longed for an idealized English countryside. They had to contend with the smog and soot and stench that spared no one in London. Nature seemed better.

---

go preach to the animals, who would presumably be less disturbed by public nudity. Of course, he did put clothes on again later – he wasn't naked when he struck the bargain with the wolf.

Our ideas about nature are often drawn from Romanticism, from the writings of men of means who longed for an idealized English countryside (an English countryside that, of course, had already been shaped by human cultivation for thousands of years). They had to contend with the smog and soot and stench that spared no one in London. Nature seemed better. It was also free, they imagined, of the political machinations and intrigues and elaborate etiquette of the city. They believed that rural people lived simple, uncomplicated lives free of the influences of modernity. But those Romantic writers and artists who did go "back to the land" tended not to fully do the work of wrenching a living from it.

The nineteenth-century British yeoman, who had to deal with this fetishizing of his toil, was no medieval serf, let alone one of the first farmers ten

thousand years earlier, whose agricultural villages rich Victorian men were beginning to unearth. He had all the advantages of the Industrial Revolution, which had been as much a revolution in agricultural technology as anything else. He was growing food for the 78 percent of the British population who no longer lived this "natural" lifestyle on the land, a population that had tripled over the century between 1750 and 1850.

And the last wolf in England was killed in the fourteenth century.

In Genesis God says "Rule over every living thing that moves on the earth" (1:28). Much ink has been spilled over how a Christian theology of dominion, derived from this verse, has led to widespread destruction of animal and plant life. But control over nature is what facilitates modern life and our ability to enjoy pleasant strolls in the woods, not to mention not dying from bacteria and viruses that would otherwise deliver pestilence of biblical proportions.

THERE ARE NUMEROUS DOWNSIDES to our modern urban lifestyle. We are surrounded by screeching cars and trains. We breathe in exhaust and chemical fumes. We need supplements and expensive lamps to deal with how little sunshine we get. Running from one thing to the next, we eat processed foods with low nutritional value. But we (rural people included) benefit from the developments of the industrialized world, like clean water and warm houses and antibiotics. There are good things about the unnatural world that we live in.

Back in 1675, King Charles II told architect Christopher Wren that Saint Paul's Cathedral was "very artificial, proper, and useful"; by "artificial" he meant that it had been designed with art and skill. The next time you're enjoying nature, remember that the park you're in is also, to some degree, artificial. And like Saint Paul's Cathedral, that's not a bad thing.

But if you see a coyote . . . stop. It's only natural. ➤

Wolf feeding, northern Europe.

CLARE COFFEY

Dandelions: An Apology

The dandelion always comes back.

SPRING AND EARLY SUMMER have always been, to me, powerfully associated with the color yellow.

First comes the forsythia, whose cut branches can be forced into bloom in the warmth of a home before the frost and rawness of winter dissipate. Forsythia is the pioneer, the first promise; later the waving daffodils that Wordsworth went on about.

In the Pacific Northwest, Oregon grape, which ends the summer with dark, glossy, spiked leaves and useful powdery-blue berries, begins dramatically differently, in a cloud of bright yellow blossoms.

The forsythia is yellow, the daffodils are yellow, Oregon grape blooms yellow, the bees darting among them are banded with yellow. Yellow, too, tints the cream produced on spring's newly lush, abundant pastures. Every week, a farmer with a Jersey cow drops off a pint of cream and a gallon of milk at my house. The cream-topped milk is not the blinding white we usually associate with a frothy glass of our daily recommended vitamin D. There is a soft, muted, but rich tint of gold, as if a lingering imprint of the sun that fed the pastures that fed the cow is reflected in the milk that feeds me. It's this reflective quality, this shadow made of light, that gives its name to the buttercup, another of spring's yellow flowers: hold it under your chin, and if you like butter (so the saying goes) it will cast a faint gleam upwards.

Buttercups can also be the impossibly vibrant color of the first, startling lumps of butterfat. When I make butter from my weekly cream (the best butter of the year, as the spring grass grows fast and sweet) nothing happens for the first interminable ten minutes or so. With the melancholy of the alchemist, I resign myself to nothing more exciting than whipped cream on my baked potatoes. But suddenly, yellow lumps appear. As I blend and blend, then knead and squeeze, watching the fatty solids slowly separate themselves from the liquid,

the pale gold of the cream grows stronger and richer. When I pour off the buttermilk, it's as if I am unveiling a molten core.

Robert Frost's famous line "Nature's first green is gold" has been interpreted in many ways: as a thought about innocence, youth, hope. I think it's about butter.

But unlike in Frost's poignant poem, in springtime, nature's first green both comes from gold and returns to gold. The sun that feeds the pastures that tint the cream recreates its own image in the big orange-yellow yolks of eggs, laid in profusion by the scratching chickens and nibbling ducks at this time of year. It is in the "beestings" colostrum that feeds new calves. It is in the new Yukon potatoes that slept in the earth all winter. It dots the green pastures with flowers. It is the chick hatching out of the egg. Yellow is everywhere in the spring and summer, because yellow is the color of new life, of life in abundance.

We greet this symphony of yellow, for the most part, with open arms. But there is one element of the yearly turn to gold that does not spark gladness in the popular heart.

It is the dandelion.

The dandelion's name, from the French *dent-de-lion*, comes from its serrated leaf, which calls to mind a lion's tooth. But something about the whole plant is leonine: not only the sharp-edged foliage, but the long yellow petals that form clusters like shaggy heads, and the lion-hearted moxie with which it shows up, uncared for, unwanted, year after year, all over the world.

The dandelion is not considered one of the glories of spring. It is a weed, a pest, a noxious invader, an enemy. It is, along with crabgrass, the foremost enemy of the manicured lawn. There is some reason for this. Dandelions are incredibly hardy, due to their deep taproots, and spread easily via wind over a wide range. You do not want these tough little urchins competing with some

Clare Coffey is a writer living in Idaho.

Sally Winter, *Time Lord*, etching with aquatint, 2022.

more delicate garden princess that you are trying to gently shelter and pamper to maturity. Just the other day, I saw one growing a little too near to one of my camas lilies. I ripped it up by the root without regret and left it lying there as a warning to its friends.

Is summer really the time for finding more reasons to work? No. It is the time to receive gratefully the abundance of creation, the symphony of yellow. It is the moment to lie in the hammock, sipping dandelion mead.

But the wholesale hatred of dandelions is not usually driven by their discrete incursions against any particular herbaceous border. The dandelions are summertime enemy number one insofar as they appear in the lawn: a space that is by definition void of flowerbeds and complex cultivation. Much ink has justly been spilled on the paradoxes of the American front lawn: how maintaining wide swaths of turf grass originated in pastoral societies in wet climates and became a status symbol in climates with neither sheep to graze nor rain to water; how the sweeping manorial expanse of green became the pocket-sized yard defended with nearly as much feudal vigor.

The transformation is charming, perhaps even noble: it tracks with an American instinct that all people are entitled to something of their very own, something to cherish and build and bequeath, that the working man's cottage is as important and dignified as the rich man's castle. But it is hard not to feel that, in the random process of cultural entrenchment, that instinct has become perverted by being squandered on an unworthy object. The lawn is a deceptive dead end, a false door in a

pharaoh's tomb, a cartoon tunnel painted by Wile E. Coyote. It is one of our most irrigated crops and yet it produces nothing. It can demand finicky maintenance, a rhythm of care around which an entire masculine persona can be built, and yet the Eden all these Adams make is mostly a green wasteland, an absence. The lawn does not require wise harmonizing of disparate forms of life, the kind of true rule worthy of the suburban king. It yields no fruitfulness, in color and form, in food, or at the level of the soil. It demands water and an endless campaign of suppression. Even as a blank space for human activity, the lawn is a failure: people are much more likely to play football or eat dinner in their backyard than in their front, since the very flatness and emptiness of the front lawn creates a kind of merciless exposure to the street.

There is no place for the dandelion in this aspirationally sterile monoculture, which is a shame. Dandelions take almost no work to cultivate, and yet they return an abundance: not only of beauty – of cheerful yellow manes that break up a flat expanse of turf and gossamer seed heads that provide playful wish-magic to children – but in food. Every part of the dandelion is useful. The roots can be roasted into a bitter herbal coffee, pickled, eaten young like a radish, or infused into a digestive tincture. The leaves can be dried as an herbal tea that supports the liver. The fresh greens are loaded with vitamins. When young, they are a delicate, spicy alternative to the more widely sought-after arugula: eat them in a salad, tossed with a mustard vinaigrette underneath soft-boiled duck eggs. The older, more bitter, tougher leaves can be slow-braised with pork belly, or chopped and fermented into a condiment.

The best part of the dandelion, though, in my opinion, is the leonine head. Picked, destemmed, infused in water and left to ferment with raw honey, they become something magical: a dandelion mead that is part tonic, part wine, all effervescent summertime joy and bottled sunshine.

The utility of the dandelion, of course, depends on whether its host lawn has been sprayed with the toxic herbicides sold specifically to combat them. In fact, companies such as Monsanto largely enabled the current state of total war on dandelions, by developing innovative products such as Roundup that allow homeowners to selectively spray weeds without killing grass. Roundup's website currently reminds you that killing invasive weeds (like dandelions) is an important act of public service: "Invasive weeds can take over an area and displace native plants, which can have negative ecological consequences. In some cases, invasive weeds can even threaten entire ecosystems."

It is ironic that Roundup has positioned itself as a tool in the service of native plants, since native plants are usually touted as the antithesis to lawns, the paradise we could achieve if we escaped the brainwashing of Big Turf. Rip up your lawn, goes the advice, kill your grass, and in the spring plant a meadow of wildflowers, beds of shrubs and forbs, elderberry and chokecherry and echinacea, perhaps even a small pond, and see monarch butterflies and native bees and all manner of birds and insects, field mice, and frogs, enjoying the bounty of a local ecosystem flourishing under the summer sun.

It is certainly a beautiful picture. I cannot say a word against it. But it makes me a bit melancholy that even among the anti-lawn crusaders, the useful, spunky, beautiful dandelion is something between an afterthought and a nuisance.

Besides, the native plant garden paradise takes a lot of work: successfully killing your turf, planning, purchasing, planting, mulching, weeding, experimenting with what works and what doesn't or paying for the advice of experts. It is good work. But many people with lawns are not looking for more work: they enjoy lawn work precisely because it feels like a sort of minimum manageable standard of care for their environment.

To these people, then, I would make my pitch: just do a little less. You can have a meadow, green and gold, dotted with the ease and abundance of summer, a feast for the eyes and the body and the bees, all by working a little less. And is summer really the time for finding more reasons to work?

No. It is the time to receive gratefully the abundance of creation, the symphony of yellow. The planting-out time of early spring is behind you, the harvest time is coming. Now is the moment to lie in the hammock sipping dandelion mead.

I do not know if this will convince anyone. I think perhaps the dandelion requires a poet, not an essayist, to make its case. Walt Whitman's "The First Dandelion" has never attained quite

---

"The dandelion holds itself no longer for its own keeping, only as something to be given; a breath does the rest, turning the 'readiness to will' into the 'performance.'"
—Lilias Trotter

---

the universal recognition of other odes to spring flowers, perhaps from a misapprehension of the dandelion as a gentle innocent rather than a brave scrapper:

> Simple and fresh and fair from winter's close
>   emerging,
> As if no artifice of fashion, business, politics, had
>   ever been,
> Forth from its sunny nook of shelter'd grass—
>   innocent, golden,
> calm as the dawn,
> The spring's first dandelion shows its trustful face.

But if the dandelion's Wordsworth has yet to appear, it has already been the occasion for meditation. Lilias Trotter, a missionary, writer, and artist in the nineteenth and early twentieth centuries, wrote:

> This dandelion has long ago surrendered its golden petals, and has reached its crowning stage of dying – the delicate seed-globe must break up

now – it gives and gives till it has nothing left.

What a revolution would come over the world – the world of starving bodies at home, the world of starving souls abroad – if something like this were the standard of giving; if God's people ventured on "making themselves poor" as Jesus did, for the sake of the need around; if the "I" – "me" – "mine" were *practically* delivered up, no longer to be recognized when they clash with those needs.

The hour of this new dying is clearly defined to the dandelion globe; it is marked by detachment. There is no sense of wrenching; it stands ready, holding up its little life, not knowing when or where or how the wind that bloweth where it listeth may carry it away. It holds itself no longer for its own keeping, only as something to be given; a breath does the rest, turning the "readiness to will" into the "performance" (2 Cor. 8:11). And to a soul that through "deaths oft" has been brought to this point, even acts that look as if they *must* involve an effort become something natural, spontaneous, full of a "heavenly involuntariness," so simply are they the outcome of the indwelling love of Christ.

The dandelion is an obvious sign of the love of Christ: as Trotter notes, for its uncalculating, diffusive self-emptying; for the artist Raphael, its bitterness and spikes serve as a symbol of the Passion. Perhaps, too, its many powers of health and healing to the body, and the evocation of the Lion of Judah in its name, its teeth, its fierce little head, invite the comparison.

And perhaps the very scorn we heap on dandelions is for this reason another hopeful sign. Welcome or not, judged as weed or flower, tormented with herbicides or cradled in the hands of children, the dandelion always comes back. It appears every year under the summer sun, growing everywhere, indifferent to our protestations, offering itself as nourishment and joy for all who care to taste its bounty. ≫

Let Them Grow
—For my grandsons

to someday stumble on the circumstance
of love, lie back, and in the quiet, try
to give shy thanks for all its providence.
I pray they'll need not know such things as, by
its sound, the caliber of ordnance.

ROBERT W. CRAWFORD

Elicia Edijanto, *In the Wind*, graphite on paper, 2022.

Promised Land

A newcomer to Saskatchewan searches for the biblical promise.

DANIEL J. D. STULAC

I AM STANDING IN A GRAVEYARD at the edge of town. The calendar reads November 1, but winter has already arrived in earnest. I scan the horizon, hoping to glimpse a ripple or a bump – any aberration – in the desolate expanse I now call home. My pupils widen, but the inescapable currents of air curl behind my glasses and I must look away. Infinite blue overhead; endless snow underfoot. Like radio static, ice crystals blow through the remnants of last summer's wheat. Wind, wind, wind. It seems to scour the whole world down to white slate.

Why on earth am I here? I am certainly not *from* here. Vocation, I tell myself. The mission of the Christian college where I now teach the Old Testament. Desperation, more like it. I have not chosen to relocate to a tiny town on the Canadian prairie inasmuch as my culture has relocated me. The tenets of modern industrialism conspire to uproot me, to "de-place" me, to keep me sliding around the planet as if it were a greased ball bearing. Twenty-five years of reading Wendell Berry and I'm still just another cog in the machine I purport to resist. I've moved (yet again) to take a white-collar job in academia. New people, new laws, new culture, new soil, new climate – all for a job. At forty-five, I am running out of time. Rooting in a place can take decades, and I have only so many decades to spare. Perhaps I should interpret Saskatchewan's mind-numbing flatness as karma for my obvious hypocrisy.

Highway signage welcomes newcomers to Saskatchewan with the slogan: "Land of Living Skies." "Don't look down," the marketing specialists in Regina seemed to whisper in my ear as I first lumbered across the border in a twenty-six-foot-long U-Haul. "You'll be disappointed if you do." Spend your money. Build the economy. Invest in potash, dirty oil, or big ag. Probably best, however, if you flick on the TV at night and imagine that you are anywhere but here.

How does one settle down while looking only up? Where's the promise in a land like this?

Old Testament Studies

Some say that Christianity licenses the unbridled exploitation of the planet, that it promotes apathy and disinterest in the material world in a way that other, earth-friendly, place-based religions do not. Christianity, they say, focuses the believer on the great hereafter, encouraging him or her to think only about leaving our space rock behind by escaping into heaven, rather than cultivating deep attention to the soil and land underfoot. The religion's inherent portability – its capacity to spread anywhere in the world – suggests that in general Christians are less attached to place.

These claims have attracted many rebuttals, including my own. We need a better understanding of scripture, a more nuanced approach to history, and fewer unfounded assumptions. I am right, I tell myself. And yet my arguments remain intellectual. When do the principles I promulgate finally move from my brain to my body, from thought into action? When will I practice what I preach?

Today I am leading another session of BLST 423 Advanced Hermeneutics. My deep background on the subject of Promised Land feels to me as solid as a rusty weathervane bending before a stiff prairie gale. As my students and I wander through modern questions concerning a literal Exodus from Egypt and the subsequent conquest of Canaan, I lead them to the New Testament. Israel's inhabitation of Canaan, I explain confidently, points beyond itself. Its material historicity alone does not account for the text's theological horizon. Books of the Bible such as Isaiah gesture in precisely this direction, and so provide a rationale for the spiritualizing tendency that seems to define Christianity's appropriation of Old Testament scripture and from which it may never escape: Don't worry about the past; look to the future! (Isa. 43:18–19). Try telling that to a modern citizen of Israel, where archaeology attracts hordes of volunteers who pay for the privilege of touching the land's recently excavated past. In comparison, my faith goes anywhere, like a feather on the breeze. My religion moves.

If the Exodus amounts to a literary paradigm, and the conquest of Canaan is a theological metaphor, does anything remain of the Promised Land as a material place? Perhaps industrialism so thoroughly uproots us because Christianity first engineered a radical divorce between people and the earth. Is the gospel intrinsically, unavoidably, landless? After all, I am here in Saskatchewan,

Daniel J. D. Stulac is an associate professor of the Old Testament at Briercrest College in Caronport, Saskatchewan.

Previous spread: Rural Saskatchewan road.

living on stolen territory, because I heard the voice of Jesus calling.

Prairie Dust

Not long after our arrival in Saskatchewan, my wife and I became regular attenders at the local Anglican church. Amid prayer requests and news pertaining to our community of white Canadians and recent immigrants from around the world, St. Aidan's weekly bulletin informs me that I worship "on Treaty 4 Territory, the traditional lands of the Cree, Ojibwe, Saulteaux, and Dakota." While entirely supportive of the point, I remain unsure what the virtue signaled by these words accomplishes on the ground itself. Does the reminder somehow help our community to become more attentive to this place than we might otherwise be? I see little evidence that such statements slow down the invasion. If a new homeowner on the edge of town, carving out a little more of the surrounding prairie as his own, were to appear in the pews one Sunday morning, we would welcome that parishioner with open arms and no questions asked. After all, Jesus taught us that sinners should not cast stones (John 8:7).

Becoming rooted in a place is tricky business, now more than ever in a world supercharged with racial awareness, and on geography defined by centuries of colonization. Ubiquitous shame has made us a continent of Pharisees who seem no nearer to solving our problems than the sinners who preceded us. I have no recourse but to turn my eyes downward, to the earth underfoot. I will begin again, at square one. I will begin with earth, for I was made from dust, and to dust I shall return (Gen. 3:19).

Saskatchewan's aridity surprises me. The air wicks my garden dry, and I find myself drawing

Storm clouds over a Saskatchewan country church.

from the roof-water collection tanks even as the growing season gets underway. The soil here feels powdery to the touch, like dry meringue. I am more familiar with gritty, East Coast custard. Fieldstones, meanwhile, seem to be as rare as arrowheads. My shovel cuts an easy furrow. The scuffle hoe slips effortlessly between the starts. No wonder the pioneers assumed they had reached the Promised Land!

When our chores are done, my daughters and I wander past the unfenced property line to fly a kite in the public meadow beyond. The activity requires so little effort that the tie-dyed object dancing above our heads begins to bore me almost immediately. As the girls take turns holding the spool, I sit down behind them on a bed of unfamiliar vegetation. Back in New Hampshire, I would know all the names: white pine, sugar maple, red oak, paper birch, and balsam fir. In North Carolina, I became a pro: redbud, beech, dogwood, hickory, and loblolly pine. Casting me upon the plains, Christian mission has robbed me of my local knowledge and expertise. Surely the plants here – bluegrass, switchgrass, Indian-grass – all have unique personalities, but I cannot tell them apart. Jesus has me starting over again as a child: feet buried in grasses I do not yet understand, eyes fixed on wind because I have nowhere else to look.

Torah Class

The land is a gift (Deut. 4:1). It cannot be grasped; it can only be received. It's not a territory to be conquered and claimed. Not a resource to be excavated and exploited. Not a virgin to be defiled and discarded. Israel inhabits the Promised Land only by adopting a posture of sheer, unadulterated dependence, by relying on daily deposits of heavenly bread. Trudging down dry wadis and out over pebble-strewn expanses, God's people feel

Canada geese fly over a southern Saskatchewan grain field.

the desert in every cracked heel and in every open blister. They do not simply pass through it; they *undergo* it, and those forty years – approximately the length of my own life thus far – remain a crucial component of the homecoming they eventually enjoy. Tattered sandals scraping over dust and gravel; eyes trained to a pillar of cloud by day and a dazzling sunset every night (Exod. 13:21). When I trekked the Negev for myself, as a graduate student, I left my camera on its panoramic setting. There too is a land of living skies.

Why do the Israelites find the Promised Land so difficult to reach? My students wonder aloud how anyone in their right mind could fail so miserably to obey God after witnessing the Red Sea crossing in Exodus 14, or after eating manna and drinking from miraculous streams along the way. Why don't the Israelites "get it"? After two or three weeks devoted to the text, the story cannot help but read as a farce, as a caricature of Israel's quest to differentiate itself from its patient but holy God. And yet I wonder if the plot is really so implausible. The historical parameters may have changed, but I too find it strangely difficult to subsist on daily bread, and to collect only as much as I need. I too have trouble accepting gifts. Something about rooting requires me, ironically, to turn up my eyes, from the soil below to the embers glowing on the horizon beyond.

Prairie Pests

Nothing drills home creation's fallen nature like an infestation of flea beetles. They invade my garden in numbers I had previously reserved only for a biblical plague. I try pinching. I try pepper. I try prayer. Nothing works. In less than forty-eight hours, the tiny insects decimate my kale and eat my arugula sprouts down to tiny nubs. I once fancied myself an agricultural educator, but here on the prairie I discover what a novice I really am. After the damage has already been done, I swallow my pride and ask around at school and at church. Longtime residents reply to my inquiry with

loving smirks. Oh yes, I am told, it's far too late in the season for kale.

Give us this day our daily salad, Lord. In the meantime, I will scratch a living from the thorns and the thistles, the hornworms and the cabbage moths. I suppose the ground is cursed (Gen. 3:17–19), but I will try again next year anyway.

Something about rooting requires me, ironically, to turn up my eyes, from the soil below to the embers glowing on the horizon beyond.

And next time, I will ask my neighbors first. I will respond to the place instead of insisting that the place respond to me.

The Acts of the Apostles

How does the Christian story train Jesus' disciples to live in – and as – creation? What do soil and land mean in the Christian imagination?

Certainly the Torah's idea of Promised Land, insofar as it embodies Sabbath rest, amounts to something more than a material, geographical, and historical category. But the fact that Promised Land points beyond itself – that it looks toward a spiritual and an eschatological horizon – does not mean that we must swap out the tangible for an abstraction. In the Bible, the material and the spiritual are not mutually exclusive categories. The new heavens and the new earth of Isaiah 65, for example, do not unfold in an alternate universe. Rather, Israel's hope takes shape through the renewal of *this world,* a world in which human beings continue to work, build, eat, play, and love. In other words, the biblical idea of land does not mutate through Christ into better psychology and an improved attitude. Rather, it encompasses both what is seen, touched, and felt and that which remains unseen, anticipated, and inferred. No,

Christians do not await cloud nine. We await, in the words of the Apostles' Creed, the "resurrection of the body."

If Matthew, Mark, Luke, and John proclaim the good news of Jesus Christ, the Book of Acts can be thought of as proclaiming the good news of the Holy Spirit. It tells the story of the church's explosion into the world outside Jerusalem. Crucially, the people who undertake missionary

---

The biblical idea of land does not mutate through Christ into better psychology and an improved attitude. Christians do not await cloud nine. We await, in the words of the Apostles' Creed, the "resurrection of the body."

---

ventures in this text do so not as conquerors but as servants of the Wind that precedes them, as if following a pillar of cloud and fire from Judea to Antioch and into the Gentile world beyond. An angel guides Philip (Acts 8:26) into an unexpected conversation, and so the gospel goes to Ethiopia. Another angel, in the port city of Caesarea, directs Cornelius to contact Peter (Acts 10:3–6). The same Wind prevents Paul from preaching in Phrygia and Galatia, but opens a door into Macedonia, and so the gospel flows freely into Europe (Acts 16:6–10). In all these cases, the Spirit goes first, and likewise, in all these cases, the Christians involved proclaim the resurrection of the body, not a detachment of the soul.

The Book of Acts suggests that Abraham's children can pop up just about anywhere. Among Jews, certainly, but also among Greeks. White or black, slave or free, Native American or African immigrant – this unpredictable Wind does not discriminate. In the same way, the Book of Acts implies that Abraham's home – God's material

promise to Israel – may be found from pole to pole. Whether a person lives in a South American rainforest, or on the Mongolian steppe, or in the outback of central Australia, rooting in a place remains an ever-present possibility, for in Christ all land is holy land. All land is a promise. All land is a gift.

Does Christianity sever humans' relationship with the earth? On the contrary, the Book of Acts – the Gospel of the Holy Wind – is "landsome" to its core. Jesus does not "de-place" me; rather, he "all-places" me – if and only if I remain dependent on the One who feeds me daily bread. If and only if I remember to look up.

Prairie Skies

Fall returns and, in an El Niño year, persists longer than it should. I am simultaneously distressed about global climate change and grateful for the unseasonable warmth. Hypocrisy never goes away entirely in a broken world.

Saskatchewan reveals a surprisingly textured landscape if one knows where to look. I have begun to explore the Coteau Hills on the southern horizon, as well as the shallow creek beds and forest pockets outside of town. A leap over the barbed-wire fence just beyond the dump leads to a game trail. Twenty gray partridges explode into view, wings beating like those of hummingbirds. They are immigrants, like me. Perhaps that is why they remain so skittish, foraging in their home away from home. I follow the path, hoping to see one of the coyotes that keep me awake at night. Maybe a badger, or a fox. Some twenty minutes later, I find myself standing at the edge of a small bluff, looking down on an aspen grove beside a pond dotted with teals, shovelers, pintails, canvasbacks, and mallards. A great blue heron stands among the rushes at water's edge. A harrier flies low over the field beyond. Wind funnels up through the coulee, whipping the long prairie grasses back and forth like ocean whitecaps. A Swainson's hawk rides the draft and disappears

over the hill, as if purely for pleasure.

A few weeks later I am sitting on my deck, watching my daughters climb the apple tree in our backyard. Crisp autumn air calls for sturdy jackets, but, knowing that deep winter will soon arrive, we milk the season for every drop of outdoor fun. Before long, I notice something brewing on the horizon and worry that the wildfires in British Columbia and Alberta will soon drive us all indoors. The "smoke" grows larger – massive, flowing, fluctuating, airborne currents, as if the sky's own fabric were folding over on itself. "Girls," I say, "come down from the tree and have a look." They register adventure in my voice and quickly follow my lead. Before we know it, a multitude of the heavenly host appears above us, praising God and saying . . .

"Birds!" cries Abigail, and she is right. Canada geese and snow geese, in mixed flocks of a size I had not imagined possible. Not hundreds, not thousands, but tens of thousands, even hundreds of thousands. They ripple by in wave after wave after wave. I could never grow bored watching this. "The sky," says Abigail, gasping. "It's . . . it's . . . alive!"

One year has passed, and again I find myself standing in the graveyard at the edge of town. Winter has returned, and I am bundled to the brim. I scan the horizon, waiting for the full moon to rise into the purple canvas above. As my pupils widen, the breeze curls behind my glasses as usual, but I do not look away. Like children laughing, ice crystals blow through the remnants of last summer's wheat. Wind, wind, always wind. I hold my arms aloft, drawn by the Spirit. My toes grip the earth below. I am dust. I am soil. I am made of daily bread. I am material creation. And I am convinced that here, too, is Promised Land. ➤

A flock of snow geese in Saskatchewan.

A Wilderness God

In the Holy Land, the desert is a place of hope.

TIMOTHY J. KEIDERLING

HALF AN HOUR'S DRIVE from central Jerusalem is a deep wadi with a hiking trail going through it. The streambed runs in a cleft between sandstone cliffs so sheer and close that you can't tell whether the ravine is twenty yards deep or a hundred. Although the trailhead is less than a mile from the main drag between Jericho and Jerusalem, the high walls block out the noise. It feels like stepping back in time, as if at any minute one might encounter a burning bush.

Even in a small and densely populated country, wilderness isn't impossible to find. In Nahal Amud, a spot near the Sea of Galilee where the Israel National Trail snakes between two cliffs, you can walk and walk and feel like you are completely alone in the world. All you hear are the birds overhead, the bats squeaking in their caves, and the rustle and grunt of what might be a wild pig.

At a time like this, when the Holy Land is torn by war and people feel unbearably far from God in their pain and rage, it seems like the land and its people are in a wilderness. There is grief at injustice, resolve for vengeance, and determination to set things right, but underneath that, for all but the strongest optimists, lies a feeling of hopelessness. Will there ever be peace? Will the two peoples who lay claim to this land ever be able to live alongside each other?

Of course, a wilderness is a hard place to be. But could it be that there is hope even here? So often, when God wants to speak to people, he does so in the wilderness, and often that's where people go to hear God's voice without the interruption

---

Timothy J. Keiderling is a member of the Bruderhof. He and his family recently returned from Israel, where he spent two years as a PhD student. He lives with his wife and two daughters in upstate New York, where he is finishing his dissertation.

of others. Why is it that God speaks to people in the wilderness? What might God be saying in the wilderness today?

In Genesis 16, when Abraham and Sarah's slave Hagar flees into the desert to escape her mistress's mistreatment, she encounters the angel of the Lord and draws courage and strength from the knowledge that her unborn son will be blessed. We read that the angel of the Lord finds her "by a spring of water in the wilderness." She asks, "Have I really seen God and remained alive after seeing him?" But more importantly, she realizes that God has seen her in her pain. After that encounter, she names God "El-Roi," the God who sees. Alone, away from people, she can both see God and be seen by him.

Moses was in the wilderness when he was tending his father-in-law Jethro's sheep and met God in a fiery bush at Mount Horeb (Exod. 3). It was in the wilderness that Moses led the people of Israel out of Egypt and again encountered God's presence on the mountaintop (Exod. 19). At different times and places, God's people have drawn new strength from direct contact with God in the wilderness. Elijah returned to Horeb to meet God (1 Kings 19). Jesus went up the mountain by himself to pray (Matt. 14:23).

Somehow, the wilderness seems to be the place to find God again. Perhaps today God will once again lead the land and its peoples through their wilderness, so that they come out the other side in a new promised land, where these verses become a reality: "As the mountains surround Jerusalem, so the Lord surrounds his people" (Ps. 125:2); and "You shall also love the stranger, for you were strangers in the land of Egypt" (Deut. 10:19); and "You shall love your neighbor as yourself" (Lev. 19:18).

John Singer Sargent, *From Jerusalem*, mixed media on wove paper, 1905–6.

The Language of the Flowers

The plants can talk. Are we listening?

WILLIAM THOMAS OKIE

*A*S A BOY I used to fall asleep at night listening to a cassette tape of Shel Silverstein reading *Where the Sidewalk Ends*. Silverstein's voice, accentuated by the cassette player's tinny speakers and whispering gears, was high and slightly rough, but playful and welcoming, a gravel country road kind of voice. My favorite tracks at the time were the ones that displayed Silverstein's bizarre sense of humor: the sharp-toothed snail inside your nose ready to bite off your finger, the sadistic dentist pulling the teeth of a crocodile, the silly king's jaws stuck together with an "extra sticky peanut butter sandwich." I didn't think much of the plaintive "Forgotten Language," in which Silverstein, accompanied by a few acoustic guitar chords and a synthesizer emulating the descent of a "falling, dying flake of snow," recalls:

> Once, I spoke the language of the flowers.
> How did it go?
> How did it go?

Silverstein's poem was lost on me in part because children are not often nostalgic for childhood. But also: I was living it. I loved plants. Like many of my friends I spent time at Nola Brantley Memorial Library, but I checked out as many books

All artwork from Julia Whitney Barnes's *Nocturnal Nature* series, watercolor, gouache, and cyanotype on cotton paper, 2020–2023.

from the adult nonfiction shelves as I did from the children's section – titles like Mel Bartholomew's *Square Foot Gardening*, or *Jerry Baker's Flowering Garden*. I joined a seed exchange I'd discovered in a magazine that, for reasons mysterious to me, was called *Reminisce*, and gained some elderly lady pen pals as a result. At one point in middle school, my garden portfolio included a vegetable garden; an herb garden; a sunny perennial border with chrysanthemums and false indigo; a shade garden with ferns and hostas; and – thanks to the herculean efforts of my slightly reluctant father in the red-clay subsoil of our backyard – a water garden: a tiny pond with orange and white shubunkin fish, a trickling waterfall, waterlilies, yellow irises, and at least one leech. (I was an ambitious but not very knowledgeable collector of aquatic wildlife.) In addition to bloodsucking worms, I also collected cuttings and seeds and other natural ephemera. My mother sewed me a vest with pockets and straps to hold the pill bottles I used for collecting seeds. And I wore it. I was a weird kid.

The language of the flowers – how *did* it go? That is what I want to know too. It seemed easy enough as a child. It is also, I have recently learned, a common tongue for many humans from many places and times. I was, as a nine- or ten-year-old, a member of that majority, the folks since time out of mind who have assumed that plants could talk.

WHAT IS THE LANGUAGE of the flowers? At the most literal level, florists have long been keen to sell us on the symbology of certain species and colors. The nineteenth century, with its advances in climate-controlled growing and shipping and its tremendous expansion in rules of etiquette, saw a dramatic increase in the floral trade, and, not surprisingly, a flurry of books purporting to guide consumers on how to speak with flowers. One popular 1855 volume was called *Flora's Lexicon*. The subtitle of an 1864 volume on "The Language and Sentiment of Flowers" promised that it contained "the name of every flower to which a sentiment has been assigned." Such flower guides were essentially code manuals for senders and recipients. According to *The Language of Flowers*, published by F. Warne, a peach blossom meant "Your qualities, like your charms, are unequalled," whereas a mignonette meant "Your qualities surpass your charms" and a spindle tree said "Your charms are engraven on my heart." According to Arthur Freeling's *Flowers* (1851), a bouquet with peach blossoms, box, cypress, marigold, carnation, and lily of the valley meant: "I am your Captive, but your Stoicism drives me to Despair; give me your Love, and return me to Happiness."

It's hard to imagine a more anxiety-inducing way to appreciate the botanical world.

People have studied the language of plants for a variety of other reasons, but the most important may be simply that plants are medicine. According to a famous Cherokee origin story, a council of animals, chaired by the grubworm, was called to deal with the problem of human cruelty and injustice toward other animals. After unanimously condemning humans as guilty, the assembly held a session at which they devised the idea of disease as a just revenge. The members of the council named diseases "one after another," with the grubworm hailing "each new malady with delight" and finally falling over backward in a fit of vengeful joy. (This is why, the narrator informs us, grubworms have crawled on their backs ever since.) The animals were plotting nothing less than extermination.

But then plants, hearing of the animal conspiracy, made haste to help the humans:

William Thomas Okie is professor of history and history education at Kennesaw State University in Georgia and associate editor of the journal Agricultural History. *He is the author of* The Georgia Peach: Culture, Agriculture, and Environment in the American South *(2016).*

Each tree, shrub, and herb, down even to the grasses and mosses, agreed to furnish a remedy for some one of the diseases named, and each said: "I shall appear to help man when he calls upon me in his need." Thus did medicine originate, and the plants, every one of which has its use if we only knew it, furnish the antidote to counteract the evil wrought by the revengeful animals. . . . When the doctor is in doubt what treatment to apply for the relief of a patient, the spirit of the plant suggests to him the proper remedy.

This origin story, collected in the late nineteenth century by ethnologist James L. Mooney, has much to recommend it, and not just the fact that a *grubworm* chairs the meeting and falls on his back laughing maniacally. Modern epidemiology has confirmed that diseases do in fact jump from animals to humans. Bubonic plague, Lyme disease, malaria, sleeping sickness, *E. coli*, salmonella, swine flu, bird flu, and perhaps Covid-19 have animal vectors. Most of the "virgin soil" epidemic diseases faced by indigenous people at the time of contact exploited the lack of livestock-keeping among Native Americans, which had left their immune systems vulnerable to European pathogens. One wonders if the grubworm story pays a sort of homage to this devastating fact.

Even more intriguingly, the Cherokee story portrays plants not just as useful materials but as active agents of healing. The spirit of the plant *suggests* the proper remedy. Every plant would furnish an antidote, *if we only knew it*. In early modern Europe, the general assumption that plants were divinely ordained as cures was systematized as the "doctrine of signatures," the idea that plants were often marked or signed to indicate their purpose. Ferns, with their fiddleheads stretching out into healthy fronds, for arthritis and rheumatism. An orchid shaped like male genitalia for impotence. Saxifrage, which appears to break apart the rocks it grows upon, for kidney stones. Eyebright, with its striped petals resembling bloodshot eyes, for ocular problems. Signatures may also be related to place: the willow tree, growing in the same sort of low, wet ground thought to produce chills and fever, seemed a good treatment for the same – and actually was, since willow bark contains salicylic acid, the active agent in aspirin. Sassafras and guaiacum, New World plants with strong aromas, seemed likely to cure syphilis, also thought to originate in the New World.

This notion that plants would be "signed" with their uses is appealing, for obvious reasons. Human health is complex and mysterious, and it's nice to think that we haven't been abandoned to our deaths. Salicylic acid from willow bark might be the most famous success of this approach to medicinal plants, but it also seems that purslane's reddish fleshy stalks, which look a little like worms, are somewhat effective in "controlling intestinal parasite loads," and that eyebright drops actually can be used to treat conjunctivitis. And smell and taste have long served as signs of potential uses and dangers. Spices like garlic, allspice, oregano, cloves, and capsicums or hot peppers, as biologists Paul W. Sherman and Jennifer Billing have documented, seem generally to have antibacterial properties. Among the Matsigenka and Yora societies in the Amazon, bitter or even poisonous plants were thought to drive out disease, while aromatic infusions kept "malodorous spirits and illness-causing vapors at bay." As the ethnobotanist Bradley C. Bennett has argued, it's probably best to see signatures not as a way to find cures but as a kind of mnemonic for dissemination of useful knowledge, created *after* a plant was discovered to be useful – a way to capture what had already been learned through careful study and experimentation.

Traditional knowledge has often served as a guide for modern pharmacology. The common landscape annual rosy periwinkle (*Catharanthus roseus*) had a wide range of traditional uses across Indian Ocean societies before it became an anti-cancer wonder drug in the twentieth century. And as Gabriela Soto Laveaga documents in *Jungle Laboratories*, modern birth control pills depended in part on the ecological knowledge and labor of thousands of Mexican campesinos who harvested the wild yam called barbasco containing a component of artificial progesterone. Western science has often behaved like a buyer bargaining for a deal, disparaging the good they ultimately resell for a hefty profit. It's all superstitious

nonsense – until there's a valuable chemical component at stake.

But the inverse of this dynamic is just as problematic – a kind of credulous acceptance of traditional knowledge as good by definition. It is certainly comforting to think that "food is medicine," that an avocado's uterus-like shape signifies an ability to prevent birth defects, or that a fig's resemblance to a testicle indicates its effectiveness for increasing sperm motility (as blithely indicated by one YouTuber with more than a million followers). But plants can be powerful for good or for ill, and the doctrine of signatures can be facile and therefore dangerous. To cite just one well-known example: the flowers of birthwort (*Aristolochia clematitis*) bear some resemblance to a uterus and the plant was used to induce both menstruation and childbirth (*aristolochia* derives from the Greek for "better birthing"), but the active ingredient aristolochic acid is a proven toxin that results in fatal kidney failure.

Renaissance advocates of the doctrine of signatures also had some very bad ideas. Theophrastus Bombastus von Hohenheim, more commonly known as Paracelsus, not only had the distinction of having "bombast" for a middle name but was a noted alchemist and physiognomist. In *Concerning the Nature of Things* he pronounced unequivocal judgments of people's characters based on their facial features: a flat nose, for instance, indicated "a malignant man, false, lustful, untruthful, inconstant." Meanwhile, Giambattista della Porta's twenty-book magnum opus, appropriately titled *Natural Magick*, presented a dizzying array of the esoteric, the experimental, and the practical. His instructions on how to "make great lettuce" – dig around its roots and add ox dung and water – were reasonable enough. But he also thought that wasps generated from horse carcasses, and serpents from the hair of a menstruating woman. In the section on "Beautifying Women," his advice on how to "correct the ill scent of the armpits" was to smear them with artichoke roots.

So no, as appealing as artichoke roots in the armpits might be, I'm not recommending a return to the doctrine of signatures. It's not the medicine itself that is most intriguing, nor even the way people supposedly discovered them by careful observation. It is a bit too comfortable to imply that advocates of the doctrine of signatures were just good observers of the natural world, proto-scientists practicing a protoscientific method. They thought something far more mysterious and powerful was going on, that they were tapping into a vital substance running underneath the

Artwork by Julia Whitney Barnes.

surface of all life. In their work, the line between the spiritual and the physical is hard to trace.

But I do think we can recover something of their intensity of focus, as well as the old assumption that the world is full of hidden potency. We might discover and behold with them the significance and integrity of other beings – not least the plants that grow all around us.

A T ONE LEVEL, it is a daily absurdity how unaware of plants most of us are. I am wearing cotton jeans, writing on a page made of pine pulp and staring at a screen reinforced with wood fibers, sitting on a rock maple chair at an oak table, fueled by a breakfast of wheat and broccoli and peppercorns, and breathing air that depends on plants reliably playing their part in the oxygen cycle. A recent biomass study estimates that the earth's plants collectively weigh about 450 gigatons, or more than 80 percent of total biomass. Humans are a tiny fraction, a mere 0.06 gigatons. And yet for many of us plants are little more than scenery on the one hand and resources on the other. This is what biology educators sometimes call

"plant blindness" or "plant-awareness disparity": a failure to see, an assumption of inertness and immobility, an indifference to ecological relationships and biochemical cycles, and an ignorance of form, growth habits, coevolution, reproduction, tactility, and taste.

And yet we do forget, ignore, disdain. There's a scene toward the end of John Steinbeck's *East of Eden* when Adam Trask casually dismisses the meaningless affairs of children, only to be rebuked by his Chinese-American servant Lee. "Do you think the thoughts of people suddenly become important at a given age? Do you have sharper feelings or clearer thoughts now than when you were ten? Do you see as well, hear as well, taste as vitally?" It is "one of the great fallacies," Lee says, "that time gives much of anything but years and sadness to a man."

Adults do sometimes recover the sense of wonder, though – often, it seems, by way of sorrow. Solomon Northup, kidnapped into slavery in 1841, found surprising solace in the plantation's pleasure garden full of pomegranates, oranges, and jasmine vines. A grieving Henry David Thoreau threw himself into natural history observation

Artwork by Julia Whitney Barnes.

after his older brother, John, died of lockjaw in 1842. Frances Theodora Parsons's husband died suddenly of influenza in Europe, and she sailed home to America in January 1891 "exhausted in body and spirit" and drugged with morphine "to insure forgetfulness." Her family recommended a charitable cause; instead she returned to the woods and meadows she had loved as a child, where she began to study "every flower within reach, noting not only its exquisite and complicated structure but its chosen haunts, its time of blooming and seeding, and even the place it might have held in the minds of poets and of painters." The result was *How to Know the Wildflowers* (1893), one of the first natural history guides, which sold out in five days and remained in print for decades. I have a copy my great-grandmother used to identify the wildflowers in the woods around Devon, Pennsylvania. Rudyard Kipling wrote to Parsons of "the debt of pure pleasure that I owe to 'Wildflowers' – I have two copies – one very muddy for field use and the other for reading when I can't go out."

The list could go on. The poet Michael Longley turned to botanizing in the midst of the Troubles in 1980s Northern Ireland; on the day the ice-cream man on the Lisburn Road was inexplicably murdered he named "the wild flowers of the Burren I had seen in one day: thyme, valerian, loosestrife," and nineteen others. The beloved British gardener Monty Don, host of the BBC show *Gardener's World*, gardens because, he says, it "heals my troubled brain." In *The God of the Garden* (2021), singer-songwriter and children's fantasy author Andrew Peterson describes how trees and gardens pulled him through his own struggles with depression. Jenny Odell's surprise 2019 bestseller, *How to Do Nothing,* (and its 2023 follow-up, *Saving Time*) emerged out of a sense of always-online despair, and she found herself staring at a California buckeye branch several times a week for more than year. Journalist

Alexis Madrigal gave up on tech writing – and rediscovered the smell of tomato leaves en route to founding the Oakland Garden Club – because, as he puts it, the internet "is mostly bad now."

These are the kind of folks I've begun to affectionately call "Persons Acting Strange around Plants" or PASAPs. An artist sketching sassafras on the Carolina coast. A surveyor making ink from pokeberries. An ecologist counting broomsedge seeds in the dissected gut of a bobwhite quail. PASAPs are very useful to me as a historian trying to understand the historical roles played by plants, since the plants themselves have left few of their own records.

I've become a PASAP myself in recent years – or rather, I've renewed a PASAP membership that apparently lapsed sometime in my early twenties. I play at shooting plantain flower heads, pick goldenrod racemes, climb Persian silk trees, and fill my shirttails with dozens of tender figs growing, preposterously, by the vape shop. Plants do their planty things entirely apart from me, and yet they are endearingly accessible at the same time. They seem immune to self-doubt, unfazed by failure, and inexhaustibly purposeful.

I'm convinced we all need PASAPs and other such weirdos in our lives, those who possess what W. H. Auden once called "that eye-on-the-object look." That look is part of a child's ordinary cognitive development. Observe a baby fixated on a piece of gum stuck to the sidewalk, or a toddler collecting bits of gravel on every evening stroll, or a preschooler studying an ant mound at the playground. As an adult, I've found myself encountering the world as "a disenchanted set of defeated and exhausted objects," as philosopher Jane Bennett puts it. It's easy to despair knowledgeably about such a world, and hard to delight in it or care for it. And so I've begun collecting again, trying to pay attention, remembering, however falteringly, that there's always more going on than I can perceive. I'm listening for the plants that talk. ➤

Why We Hunt

In the woods, to be a predator is a privilege.

TIM MAENDEL

I AM TWENTY-FIVE FEET off the ground in a slightly swaying tree. It may as well be a hundred feet, since the pre-dawn darkness obscures my view to the ground. This moment of stillness caps an hour in which I feel I have already lived half a day. My alarm clock had ended my dream preview of the deer now somewhere nearby, one of which is likely having its last night. Dream fragments still swirling around me, I grabbed my bow and drove to the edge of the woods, soaking up the last heat from the vehicle before stepping out into the cold. Unseen creatures followed my approach. I briefly became disoriented as I tried to find the way to my hidden ladder. Finally, after the dangerous climb, the snap of my safety vest's clip signals the end of my

Tim Maendel lives at Bellvale, a Bruderhof in Chester, New York, with his wife, Kathleen, and their dog.

Tim Maendel heads to the woods with his bow.

intrusion. The disturbing wake of my entry melts away. I am now part of the tree, part of the forest's quiet calm, part of the natural world.

I have left the chatter of my life and have joined an ancient peace, another world where I am only a student and definitely a foreigner. But even as an intruder, I am a member of an age-old society of hunters: men and women throughout the ages united in their mission to provide food for their families. They have, like me, left their homes and kin and crossed a border into nature, into a wild ruled by animals whose ways they have to learn. Sometimes when hunting from the ground I wonder if a Lenape hunter ever stood exactly where I am, seeing the same rocks and the ancestors of the same trees, hearing the same sounds I am hearing now.

The forest soundtrack tests my discernment. Squirrels and chipmunks in the leaves briefly fool me, but there is nothing that makes my heart race like the irregular punching sound of the deer's hooves now approaching. I lean forward slightly, and there it is, antlers weaving and bobbing as it seeks acorns. When I stir, it briefly stops to read my movement. I am now in plain sight, but if I hold still my image does not register as danger, so it goes back to food-seeking. A few more steps bring it into the alignment I need for a lethal shot. Minutes later I am standing over the stretched-out deer.

I have sometimes found myself in strange disbelief at this point. The deer looks so peaceful and unharmed that I wonder if I am really responsible for this, or if it is even my deer. I kneel and grab an antler. No human has ever been this close to this deer; nobody has ever touched what I'm touching. Neither has anyone helped it find food or provided it shelter in subzero snow storms, summer downpours, or scorching sun. It has seen the deep forest secrets. The meat under its skin comes from food it found itself.

I murmur words of thanks and then start field-dressing it, the warmth of its moist innards the first gift to my frigid hands. I identify the organs as I remove them and, except for the liver and heart, leave them for the forest food chain.

MOST STATES REQUIRE YOU to take a hunter safety course before you can purchase a hunting license. Typically, these courses cover proper handling of firearms and personal safety (one of the most common hunting accidents is a fall from a tree stand), tips on approaching landowners for permission, and public relations. But more time is spent on principles of wildlife management, wildlife identification, hunting ethics, and the hunter's responsibilities toward animals and the environment. Out of respect for the animals, and to improve public perception of hunters, these often include "fair chase" rules, which trace back to customs developed in the Middle Ages to increase the challenge and integrity of the sport. In 1967 naturalist Bill Wadsworth and some fellow bowhunters started work on an education program to train hunters and protect the sport they loved. "If bowhunting as we know and enjoy it is to survive, we must be hunters who appreciate and respect the environment in which we hunt, as well as maintain a strong desire to uphold the highest standards for our sport," Wadsworth said. His curriculum is now used nationally.

Bowhunting offers special satisfaction, as it gets hunters closer to the game and puts them in touch with centuries of people who crafted their own weapons long before firearms were invented. It requires skill and patience, which add to its allure. In my bow class we spent hours learning the ways of the deer, fair chase, and humane kill. We learned internal deer anatomy and were shown acceptable shot placement, along with a list of shots never to attempt due to their high chance of wounding and causing suffering instead of a clean kill. We were advised to wait thirty to sixty minutes before setting out

to track a deer, since even a lethal bow shot will usually cause an animal to bed down as it tries to recover from the bleed. The sight of a hunter at this stage will stress the deer into using its last strength to move to a hiding place in an area of dense underbrush where it is likely to die alone, its meat wasted. We practiced tracking a blood trail the instructors had made with ketchup. As a final measure of our worthiness to be humane hunters, we had to prove our marksmanship.

Not every hunt goes as I would like it to. I am ashamed to think of times when a poorly placed shot, usually the result of over-eagerness fueled by a long wait or several unproductive hunts, did not end in a peaceful death. Once my first crossbow shot paralyzed a deer, leaving it alert but immobile. After a second shot I was surprised at its calm, so I walked up and stood close to it. At this point a clear mental image appeared, one that often comes when close to an animal or fish I have caught – the face of my dog, Sammy. Perhaps it's because he represents the closest bond to the animal world I have in my normal life. I wished he was here. But the vision faded as I saw and heard the deer's death throes. I thought of turning away but duty held me in place, and I found myself speaking aloud to the animal, apologizing for the way it turned out and thanking it for its life. I was surprised that my voice – so close and surely the first human voice that had ever spoken to it – didn't startle it. I have no idea what emotional intelligence the deer has, but I like to think my soft voice was calming it, easing its transition.

ANOTHER SIMILARITY between hunters and their prey is the pull of desire that can cause humans and creatures alike to throw caution to the wind. Out in the woods, watching this force override animal behaviors needed for safety and survival, I recall legends of boat pilots lured to their death by lovely mermaids or sirens. Or King David, who threw away his hard-earned honor at the sight of another man's wife. Similar stories of derailment can be found in almost any news feed. Here in the wild, the same force lures turkeys and deer out of firm safety at certain times of year. The most experienced and cautious buck will step out of concealing brush and dash across an open field at the sight of a doe or competing buck. Artificial scents put out by hunters or grunts imitating a male announcing his territory have the same effect.

Years ago, my sixteen-year-old son and I decided to try turkey hunting. We set up a cloth ground blind and within shooting distance a rubber jake (young male turkey) decoy standing proud and tall, sporting a short beard, paired with a rubber hen with head down as if feeding. YouTube had helped me learn the basic call of a cheeky young jake, which remarkably brought an immediate return call and then the appearance of an enormous tom turkey who needed to see what was going on. Weeks before, the same turkey would have sprinted for cover if it had seen me a thousand yards away, but in its current hormone-charged state the idea that a teenage jake had got a member of his harem was too much. It ran past our blind's window close enough to touch, literally spitting in its anger, and started pecking the head of the rubber bird with a violence that threatened to destroy the decoy. We both burst out laughing, which is probably why my first shotgun blast missed. Apparently tuning out even that thunderous sound, the tom went right on pecking. "No," said my son, "this is how you do it." Taking the gun, he finished off the bird.

THE STREAM OF FAWNS and turkey chicks entering the world every year are a fountain of continuity, but unchecked population growth would soon deplete the available food supply. The predator plays the necessary role of healthy limitation. Each species does its part in balancing overgrowth. Shrubs would overgrow fields and woods if not for harvesters such as deer, but if the deer become overpopulated they kill off the same growth. Enter the wolf and coyote. Humans

have certainly played a part in unbalancing this stability. But hunting, if practiced within limits, is one way of providing the necessary predator. It also strengthens our connection to nature and clarifies our place in the food chain.

Before modern hunting laws, humans hunted many animal species close to extinction. Deer were almost wiped out in the early 1900s, but through conservation the United States now has an estimated 36 million. Wild turkeys were taken from a national population nadir of 30,000 in the late 1930s to around 6 million today. But even this conservation success story carries lessons in over-meddling by humans. You might think that catching a few of the remaining birds and using them to breed and release more birds would have been the easiest solution. Instead, this had the opposite effect, as the bred birds were unable to survive on their own due to all the human help they had received. Instead, repeated netting of parts of flocks and moving them to less populated areas, as well as predator control, proved to be the solution. Today's hunters, who are beneficiaries of these conservation efforts, have a responsibility to support them.

For me, it's paradoxical that the hours spent in nature bring such great peace to my soul, since I come there as a predator. But in so doing, I am included in the natural cycle that is necessary for life. The animals I hunt are also hunted by animal predators, or killed by vehicles on the highway, or die of starvation if their numbers get too high. I get to experience firsthand where meat comes from and share this experience with others.

I SPENT THE LAST LIGHT of the season this year standing in the same field as a young deer that I had stalked after deciding to let it grow even though it was in easy reach. I watched it feed and then raise its head to match my stare. We stood like that for several minutes until movement to my right distracted me to an even better reward – a hunting red fox, tail almost as long as its body, came silently floating within feet of me before it turned to disappear with the daylight.

Moments like this stay with me and I revisit them when I need an escape from the noise of life. I wish every person in the world could feel this rejuvenating connection to creation. For some, a hike topped by a mountain view or time spent at a mountain stream does it. For others, it's hunting or fishing that provide this. It's a privilege that we must steward with care and pass on to the next generation, along with the respect, humility, and gratitude that mark the spirit of a true hunter. ➤

Illustrations by Alison Kleinsasser (the author's daughter).

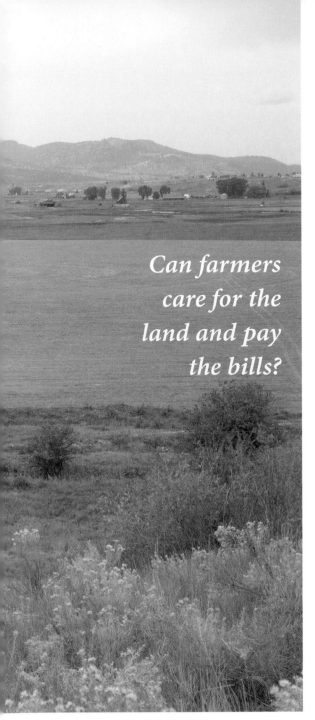

Can farmers care for the land and pay the bills?

Cory Miller's Montana ranch (see page 58).

Saving the Soil

COLIN BOLLER

IN A WORLD where industrialization and modernization often dictate the course of progress, a revolutionary shift is taking root in the heart of agriculture. The movement toward regenerative farming practices is not just about adopting the latest scientific advancements; it's a return to the wisdom of the farmers who have worked the land for generations. Farms are businesses that require financial sustainability. Better growing practices must align with economic viability, saving costs and boosting profits to ensure the survival of farms and those who run them.

Regenerative agriculture is an approach to farming that focuses on revitalizing and strengthening the health of the land and its ecosystems by prioritizing practices that promote soil health, biodiversity, and carbon sequestration while reducing the reliance on synthetic chemical pesticides and fertilizers. The economic and environmental incentives, along with the clear connection between the quality of food and our health, all point to a bright future for regenerative farming. A critical next step is helping farmers overcome the practical barriers preventing them from fully realizing this vision for farming.

The Green Revolution during the past century made many advances in farm productivity. However, it did so at the cost of the soil, severely damaging and depleting the soil's microbial ecosystem, leaving farmers dependent on carbon-intensive traditional fertilizers and government subsidies. Economic volatilities such as price increases and supply-chain shortages put farms under crippling financial strain.

We now know that prolonged use of chemicals results in the loss of biodiversity in the micro-biome, which reduces nutrient availability and causes the breakdown of soil aggregate structure at the microscopic level, leading to soil compaction and erosion and to compromised plants that are more susceptible to pests and diseases while producing nutrient-impoverished crops. As the soil deteriorates and erosion increases, runoff carries a greater amount of fertilizers into surrounding waterways. More fertilizer is required to supplement what was lost, only to wash away again. This cycle is costly and threatens the future viability of the farm.

Fortunately, there is a solution. Farmers are increasingly focusing on regenerating soil health by enhancing soil structure, fertility, and resilience. Driven by new scientific understanding of the role of microbes in the health of the soil, practices such as reducing tillage, cover cropping, crop rotations, and animal integrations are increasing in popularity. As Elaine Ingham, founder of the Soil Food Web School, puts it: "If we want clean water, we have to get the biology back in our soils. If we want to grow and harvest crops, we have to build soil and fertility with time, not destroy it. The only way to reach these endpoints is to improve the life in the soil."

A growing number of farmers are looking to rebuild the microbial ecosystem by adding beneficial microbes into the soil. These microbes can be grown using specific composting systems, such as vermicomposting, and then extracted into a liquid form that can be applied in furrow with the seed at planting or used as a seed coating, soil drench, or foliar spray. This approach can accelerate the transition toward healthier soils while providing a low cost, environmentally friendly alternative to chemical applications. While adoption of this practice is still in its infancy, with only a fraction of American farmland being actively improved though regenerative methods, the results so far have been exceptionally promising.

RICK CLARK farms seven thousand acres of soybeans, alfalfa, and grain crops in western Indiana, where he has defied the limits of what is feasible with regenerative farming at scale. He has eliminated synthetic fertilizers and reduced his diesel fuel consumption by 50 percent, which together saves him $2.7 million a year, leaving some seed and machinery costs as his only major input expenses.

Cover cropping and diversifying crops have resulted in less total yield per acre annually but more profitability – that is, a smaller harvest overall, but a better return on investment. "When you start to look at the savings that you get from the reduction or elimination of inputs, you then start to shift away from a yield-driven mindset. This is a profitability mindset," he says, also noting that his farm achieves the rare trifecta of "regenerative, organic, with no tillage."

Meanwhile, he's also bringing seed costs close to zero by developing his own supply: "The beans we harvest this year, we will clean them, and that's where we'll get our supply for next year's bean fields," he explains. This means that the five varieties of soybeans he grows are adapted to the specific land they're on, symbiotically in tune with the microbes in the biome. Further, over generations, the beans develop heritable changes to benefit future crops – not only via the evolution

---

Colin Boller is CEO of Hiwassee Products, a business providing equipment for regenerative farming. He and his wife live at Hiwassee, a Bruderhof in Madisonville, Tennessee.

of the genome but more rapidly via epigenetic responses to external conditions. In short, he expects his crop to be better suited to the land it's on with every passing year.

ACROSS THE COUNTRY from Clark, Douglas Poole is on the same wavelength. He farms wheat on fifteen thousand acres of marginal farmland with eight inches of annual rainfall in central Washington, farmland his father bought and harvested with a $15,000 combine. Today, Poole notes, the combine he would need on that land would run him a million dollars – yet neither the yield nor the price of wheat has risen enough in the intervening decades to be remotely worth that expense. This got him radically rethinking what profitability would mean for his farm – most significantly, in terms of weaning off synthetic fertilizers.

He recalls speaking with salesmen about which mix of expensive fertilizers to buy: "They were talking about different fertilizers and why you needed zinc or magnesium or copper or whatever else. But they said, well you got to be careful because if you mix too much of this, well, here's this Mulder's chart – it's going to antagonize these four things over here. You need a little bit of that, but of course, that pisses these four elements off." From a sales perspective, Mulder's chart of nutrient interactions seemed to present an opportunity to sell him countless elements and then more elements to counteract the effects of those.

But what does this major expense do to help him? Poole continues: "My biggest weed around here is a Russian thistle. What does a Russian thistle love? High nitrate. What do I do? I keep continuing to put nitrates in the soil and all I'm doing is helping the Russian thistle out. Which of course makes the Roundup salesman happy because I got to have more Roundup to control the Russian thistle. People that are around soil health know this vicious cycle. I just think there's a positive cycle out of this."

To Poole, the way out is right there under his toes – "just get out of the way of the soil biology." No added elements or chemicals, just apply

---

Rick Clark farms soybeans, alfalfa, and grain crops in Indiana.

compost and let nature do the rest. Growing plants is "biology, not chemistry," he sums up. ("And," he adds, "I don't have to write a check for it.")

TWO STATES OVER in Montana, Cory Miller also extols the benefits of soil biology. Formerly a developer of natural cleaning products, he now farms a thousand-acre ranch producing sod, hay, beef, pork, and honey. He also shares his enthusiasm for soil biology and regenerative practices on a popular YouTube channel.

He doesn't wax eloquent about going green or saving the planet, although those considerations are something he finds important. Like Poole, he goes straight to the financial benefits: "I got this place, and I had six months to figure out how to make it make money and to do that without a significant investment in infrastructure," he says. The eye-popping cost of fertilizer made him seriously consider "how to build something more

resilient that can be sustainable over the long term. Everything pointed to biological farming and letting biology release the nutrients in the soil."

He commiserates with others facing this transition. "You can waste a lot of money farming, a lot of money, and you have to set some limits on where you go to find your success."

These farmers have found their success by recommitting to a knowledge of the soil and respect for natural cycles. In embracing a profitability mindset, the implementation of regenerative practices enabled them to reclaim financial independence.

THE BENEFIT TO THE BOTTOM LINE is not the only important tenet of regenerative agriculture. For one, healthier soil makes for healthier food. As Miller puts it, "The density of nutrition that we can pack into food is the number one reason for doing it. There are a lot of bad things that are in our food supply that end up inside our bodies. The science suggests that a lot of this has to do with chemicals in the food we buy at the grocery store." He continues, "After learning about this and studying it and practicing it, I feel

Douglas Poole farms wheat in central Washington.

like a totally different person. When I was busy all the time, fast food was the option, just pack it in, eat as quick as I can, get back to work. Now I carry a bag of carrots in the tractor, and I can eat them slow."

Adam Henderson, who farms two thousand acres in Iowa with his brothers Andy and Aaron and their father Bob, elaborates on the scientific connection between soil health and nutrition. He recalls a lightbulb moment he had in a college class when his professor put a slide up "showing the degradation of the nutrient value of corn over the last forty years, how conventional corn compares to the nutrient value of GMO, the difference between starch and protein values." That stuck with him. Today, "we feed our animals a lot of what we grow, and our animals are healthier, we eat the meat out of those animals, so hopefully we'll be healthier too. That's what the whole thing is, to bring it around full circle to where animals are healthy, people are healthier, and soil is healthier."

And, last but not least, the earth is healthier as well. As caretakers of land, farmers understand the broader environmental impact of their practices. Regenerative farming can mitigate

pollution, conserve water, reduce erosion, and contribute to a more resilient ecosystem.

"Mother Nature does some crazy things," notes Miller. "She brings hurricanes and tornadoes, and all those things are for a purpose. They're changing the dynamic of what's growing." Being attuned to nature, even at this disruptive level, can help farmers "build something resilient and prepare for those bad years."

For Clark, it didn't take a hurricane to bring this point home. He remembers a time when his farm was still high-tillage and "we got a one-inch rain event. Just one inch on a one-percent-slope field. The erosion that occurred just blew me away. The field was on the road. That's when I woke up and said, 'It's time to make a change.'"

Erosion was also a deciding factor for the Henderson family. They started using compost extract, Andy explains, because "if you reduce your erosion, you're going to have more of your soil and it is going to stay put. So therefore, your nutrients are also going to stay put too. Your water-holding capacity is going to increase and that will help eventually make the crops better in a drought-stress year."

Cory Miller's ranch in Montana produces sod, hay, beef, pork, and honey.

Over the past ten years, the Hendersons cut back on tilling and made many other changes to the farm as they learned more about regenerative practices, from rotational grazing to monitoring microorganisms. When they started out, Adam recalls, "We knew from soil testing that we didn't have the biology in the soil. There's hardly any protozoa." But year over year, these results have improved, and meanwhile their respect for nature has deepened. "Don't treat the soil as just a growing medium for plants. There's more stuff going on down there than there is up top."

Bob sums up with advice from Gabe Brown, author of *Dirt to Soil*: "'I used to go out and figure out what I was going to kill that day, a bug or a weed or something. Now I try to figure out what I'm going to keep alive.' That's a huge statement because that's where we're at. We'd like to try to keep as much stuff alive as we can instead of trying to kill everything."

THOUGH PROMISING, the journey toward regenerative farming is challenging. Aaron Henderson advises, "There are a ton of risks involved in this, and that's a huge reason, a huge hurdle, why people don't just jump full step into it. Baby steps are the key. Slowly work your way into it, figure out how it stays profitable for you, and you'll slowly see the results over time."

The transition from conventional practices to regenerative methods may involve initial costs, a learning curve, and a reluctance to change. "As far as jumping headfirst into it," he continues, "it's a scary thing financially. But you can see the benefits over the years."

TO BE SURE, extracting enough beneficial biology from compost to cover a large farm is an operational constraint. Without the right equipment, the process is time-consuming and messy at best. At worst, there is a high likelihood that valuable material will be wasted and even pathogens produced and spread.

I know these challenges firsthand. When the Bruderhof made the shift to regenerative

Bob Henderson and his three sons farm two thousand acres in Iowa.

practices on our own farms, we recognized that an abundance of scientific data pointed to the benefits of rebuilding microbial ecosystems in the soil. Currently, all fourteen Bruderhof farms are either fully regenerative or transitioning toward that goal. Yet, as we began implementing regenerative methods, it became clear that there were few tools available to efficiently execute these methods at a large scale. Our solution was to create our own tools to support our farms and others looking to make the same transition.

At the new Bruderhof on the former Hiwassee College campus in Madisonville, Tennessee, we have recently launched a business endeavor focused on providing equipment solutions for regenerating soil biology. Hiwassee Products makes equipment systems that address biological soil building from composting to extraction to application and the testing of results. These products are geared toward rebuilding soil microbial communities in a way that's scalable and adaptable to various agricultural systems.

We offer equipment solutions that will work for small or large-scale farms, so both can make the shift to more regenerative practices.

Hiwassee Products works closely with farmers, integrating their requests with the advice of the soil lab technician to develop this equipment. These solutions bridge the gap between science and methodology, linking the two by making best practices proven in the lab an operationally efficient reality on the farm. Like the farmers we serve, we are still learning what works in our own fields and workshops. By drawing on the latest science alongside the wisdom of the past, we are striving to help farmers meet the demands of a rapidly changing world. As challenges loom, regenerative agriculture holds the promise of surviving and thriving. Elevating the vitality of soil is a strategic investment in the planet, promoting human health, fostering the growth of nutritious food, and ultimately yielding a profitable bottom line. ➤

Hiwassee Products's soil consultant at the controls of the Continuous Flow Bio-Extractor.

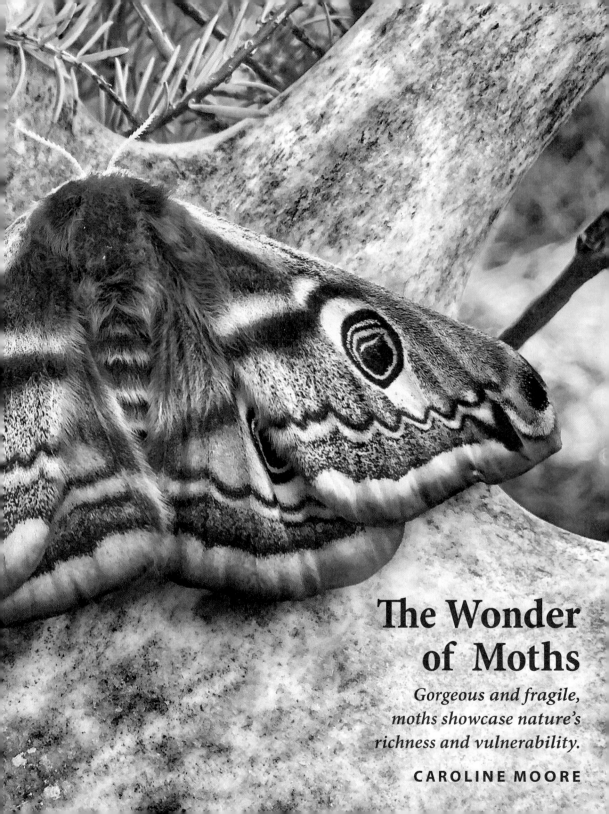

The Wonder of Moths

*Gorgeous and fragile,
moths showcase nature's
richness and vulnerability.*

CAROLINE MOORE

I HAVE BEEN WONDERING AT MOTHS since I was a small child. When I inherited my grandfather's old moth trap, twenty-six years ago, I started logging the moths in my Sussex garden, where I have so far found 903 species. Of these, seventy-five are rare enough to be categorized as Nationally Scarce, and twelve are either on the Red List of threatened and endangered species or proposed for it.

Despite its age, my ancient Robinson trap is still perhaps the best model on the market. Moths are attracted to the light of the mercury-vapor bulb, and are funneled down a cone into a metal drum beneath. Cardboard egg cartons in the drum provide sheltering, shadowed caves where the moths rest quietly until morning. I come down early, and very gently lift the lid to inspect the glistening riches within. After a good mothing night, warm and humid, there will be over a hundred different species clinging to the boxes, to be lifted out, identified, counted, and, if they're rare, photographed for the county moth recorder. The moths are then usually released unharmed, though I have to help them to dodge the greedily opportunistic blue tits and robins.

My hobby – perhaps obsession would be a better word – makes every morning thrilling. The excitement is as intense as the Christmases I remember from childhood: I never know what the trap will hold. And looking at moths keeps alive that childhood sense of pure wonder.

If the beauty of moths makes their study aesthetically rewarding, their sheer diversity makes it ecologically significant as well. There are about 2,500 species of moth in the British Isles. (Only two are likely to be the larvae found munching on your woollies – always the best cashmere, for choice, since it is more digestible.) For comparison, there are only fifty-seven species of butterfly – fifty-nine if one includes those regular migrants the painted lady and clouded yellow. Butterflies and moths, of course, are both members of the same order of insects, Lepidoptera. Butterflies, however, are a minor branch; the vast majority of the world's Lepidoptera are moths.

Lepidoptera means "scaly-winged"; every moth wing is patterned with a mosaic of microscopic scales, as fine and fragile as dust. Every minuscule scale is pigmented, and its structure both reflects and diffracts the light, creating an iridescent shimmer of extra colors like oil on water – glittering blues, golds, and silvers. The scales are easily rubbed off, but the wings of even the plainest newly hatched moth have the sheen of silk. The very name of the merveille du jour suggests its evanescent beauty, patterned with ripples of peppermint green, silver, and black. Even the less gaudy moths have a superbly subtle beauty. A yellow shell, for example, has numerous delicately scalloped lines of brown, as if drawn with a wavering fine-toothed comb, across its corn-gold wings.

The more closely you look at any natural beauty, the more you see: there is an intricate excellence to be discovered. This is especially true of the so-called micromoths – the (usually) smaller and (usually) more taxonomically primitive species which make up the bulk of the moth population. There are around nine hundred species of British macromoths: these are the families of larger moths that are found in a standard field guide. The largest of these are the hawk moths, such as the privet hawk moth, which has silvery-grey wings about five inches across and an abdomen and underwings striped black and bubble-gum pink. Another pink-tinged moth, the rosy marbled, is among the smallest of the macromoths, with a forewing measuring about three-eighths of an inch. Some micromoths are larger than this.

Caroline Moore was introduced to the study of moths by her father. She taught English literature as a Fellow of Peterhouse, Cambridge. For decades she has been recording the moths found at her home in East Sussex.

Previous spread: Emperor moth pair on a fallow deer antler. Photograph by Andy Newman.

Moth enthusiasts generally start with the eye-catching macromoths, and are then lured into the micros. It is among the micros that the most amazing diversity of form and habits are found, since their smaller size allows them to exploit a wider range of habitat niches. I have to admit that one or two families of micromoths are the equivalent of a birdwatcher's "little brown jobs": I can only identify a few of the most striking among the 109 species of Coleophoridae – pale-brown, tiny moths with long thin wings and forward-pointing antennae. Species that cannot be identified by external characteristics, however, can be separated out by "gen. det." – a rather sinister abbreviation that stands for "determination by genitalia examination." The genitalia of males and females of each species of moth have evolved to fit uniquely like keys in locks; this handily prevents interspecies hybridization.

The invaluable *Field Guide to the Micro-moths of Great Britain and Ireland* by Sterling and Parsons illustrates and describes a mere 1,033 of the easier species. Taking a good quality close-up photograph helps identification – one can enlarge the diagnostic features such as wing-pattern, labial palps, or spurs on the legs. And good photographs also reveal the extraordinary beauty of many of these miniature creatures. One of my favorites is the tiny and rare *Bisigna procerella*, a regular visitor to my trap: it is coppery-orange, patterned with a loop of black and silver, and its forewing is only about a quarter-inch in width and length. Many micromoths seem at first glance small and brown, but viewed close-up they sparkle as if stitched with metallic thread, or have the rich patterns of minute Persian carpets.

The huge diversity of size, shape, and wing patterns makes moths a useful source of

Privet hawk moth. Photograph by the author.

ecological data. As with birds, there is a range of species that are readily identifiable by list-obsessed amateurs. If we submit our data to the county recorders who are compiling data on all Britain's Lepidoptera, we can be dignified by the appellation of "citizen scientists" rather than just being "twitchers" or "nerds." But moths are even more valuable than birds to ecologists, because there are so many more species, and because they are widely recognized as important markers of biodiversity.

This is partly because of their place in the food chain. Moth larvae eat a wide range of plants: some are generalists, and their Very Hungry Caterpillars will eat many sorts of leaf; some are specialists, which does not always work to their advantage (the rare white spot, for example, will only eat a form of campion called Nottingham catchfly). Some micromoth larvae feed on fungi, algae, mosses, or lichen; others feed on decaying plant or animal materials. A cousin of the clothes moth feeds on moldy corks in wine cellars; another is found in fox excrement (this moth is horribly common in my garden). There is even one moth, the wax moth, whose larvae can consume and digest plastic bags, though its preferred food is the wax in honeycombs.

A wide range of moths indicates a wide range of plants, which they help to pollinate, and a good diversity of habitats. Moths and their larvae are consumed as well as being consumers. They are important food for birds, bats, hedgehogs, frogs, toads, lizards, and insects such as spiders, wasps, hornets, and beetles. At the height of their breeding season, great tits and blue tits in the United Kingdom feed an estimated two billion caterpillars to their nestlings *every day*. And those harassed parents have to eat too.

Yellow shell moth. Photograph by the author.

The fragility of moths makes them particularly valuable to ecologists as indicators of the health of our ecosystems. They are like canaries in a mine, the first to suffer. Moths are particularly susceptible to pollution of all kinds: water pollution, air pollution, chemical pollution, and light pollution.

The last is the least written about, but the one about which private individuals can make the most difference. Light pollution creates havoc in the natural rhythms of wildlife, disrupting patterns of migration, feeding, and reproduction. The problem of streetlights, outdoor security lights, and floodlights has paradoxically been made worse by half-hearted attempts to be "green." LED lights are now favored, on the grounds that they use less energy. LED lights, however, generally emit a bluer spectrum, which is more attractive to moths. What would be truly energy-efficient, of course, would be to turn outside lights off, or at the very least to make them motion-activated.

Water pollution affects the moths that live beside our rivers and streams. There are specialist species whose larvae live below the surface of the water in ponds, marshes, and slow-moving rivers. The larvae of the China-mark family, for example, feed on water plants and build underwater cases out of fragments of leaf.

Air pollution is devastating for moth diversity. Many moths feed upon lichens, which are among the first organisms to suffer from pollution. A lichen is made up of a fungus cohabiting in a symbiotic relationship with green algae, or cyano-bacteria, or both, which can photosynthesize the sugars that their fungus host cannot. But these composite organisms have no roots or protective skin, and they absorb moisture and air directly through their surfaces, making them highly sensitive to air pollution. Sulphur dioxide is among the worst pollutants, produced by coal-burning; nitrogen dioxide is produced by cars but also by agricultural fertilizers and livestock waste. Both dissolve in water to produce acid rain.

There are so many tales of ecological doom and gloom that I'd like to introduce a ray of hope. We don't hear so much about acid rain nowadays, for the simple reason that our air is cleaner than it was. And the impact of Britain's Clean Air Act can be traced directly in its remarkable effect on moth populations.

Most of the lichen-eating moths are micros, logged only by nerds like me, but the fortunes of the conspicuous lichen-eating footman family are readily tracked by amateur recorders, with hugely cheering results.

The family got its name from the common footman, which looks as if it is wearing a grey uniform with a yellow stripe down each side. A study published in 2019 found that its population has risen by 40 percent in the last thirty-five years. That is nothing, though, compared to the buff footman, which increased by 815 percent in the same period – it had been almost extinct in Sussex. The scarce footman, which looks like a more tightly rolled-up version of the common footman, had the most startling rise of all. The population rose by 2,035 percent – so it is not so scarce now!

Chemical pollution is still the chief killer of moths – but there has been increasing awareness of the dangers for human health in the indiscriminate use of pesticides, herbicides, and fungicides, and there are growing efforts to farm with rather than against nature, though there is a long way to go.

Urbanization is still on the rise, however; habitats are being irrevocably destroyed, and many insect populations are in swift decline. The species-rich habitats that do remain are increasingly patchy, and their populations are correspondingly vulnerable because they live on small and isolated islands in a sea of ecological desolation.

I am sometimes asked what I am "doing right" to have found so many moths in my own garden (apart from looking for them). The truth is that it is not really me – it's the landscape around my garden that is so fruitful. The light from my

trap shines out across the Dudwell Valley, which is a classic mosaic landscape – a patchwork of small fields, mercifully unsuited to intensive farming, with old hedgerows, patches of ancient woodland, boggy water meadows, and dewponds. This is a landscape rich in heterogeneity, a mix of interacting habitats. The importance of these mosaic landscapes is often overlooked in our struggle to save more specific types of endangered habitat – reed beds, woodlands, or moorlands.

A crucial factor for biodiversity is not just the variety of these patches, but their interconnectivity: even small populations have greater resilience when movement between them is possible. I am lucky here, because the River Dudwell forms a superb wildlife corridor. Rare moths that breed only on shingle beaches regularly find their way to Etchingham from

Rye, coming up the River Rother and turning left into the Dudwell. These have included such Nationally Scarce species as *Neofriseria peliella, Pediasia contaminella, Platytes alpinella, Ancylosis oblitella, Scrobipalpa ocellatella,* and *Oncocera semirubella.* I also gather a fine array of migrants: new breeding species are colonizing all the time. Lepidopterists must be the only internet group to celebrate both migrants and colonization!

Living, as I do, in an old rectory, I sometimes think of the clergymen who I imagine named some of the moths that come to my trap – the Hebrew character, the many varieties of Quaker, conformist, and nonconformist moths, gothic arches, seraphim. . . . The Bible does not have a good word to say for moths; but surely these men of God found the wonder of creation in their pursuit of lepidoptery, just as I have. ➤

Flame carpet moth. Photograph by Anthony Roberts.

Lilias Trotter, *Oleanders*, watercolor, 1893

Reading the Book of Nature

Four writers study the divine
as revealed in the natural world.

Lilias Trotter

Gerard Manley Hopkins

Augustine of Hippo

George MacDonald

Lilias Trotter

*Isabella Lilias Trotter was a British artist and a Protestant missionary
to Algeria in the late nineteenth century. She chose the mission field
above a promising art career, and never lost her love of nature.*

IT WAS IN A LITTLE WOOD in early morning. The sun was climbing up behind a steep cliff on the east, and its light was flooding nearer and nearer and then making pools among the trees. Suddenly, from a dark corner of purple-brown stems and fawny moss, there shone out a great golden star. It was just a dandelion and half withered, but it was full face to the sun, and had caught into its heart all the glory it could hold, and was shining so radiantly that the dew that lay on it still made a perfect aureole round its head. If the Sun of Righteousness has risen upon our hearts, there is an ocean of grace and love and power lying all around us, and it is ready to transfigure us, as the sunshine transfigured the dandelion, and on the same condition that we stand full face to God. Turn your soul's vision to Jesus and look and look at him, and a strange dimness will come over all that is apart from him, and the Divine "attrait" by which God's saints are made even in this twentieth century will lay hold of you. For he is worthy to have all there is to be had in the hearts that he has died to win.

Lilias Trotter, ed. Constance E. Padwick, *Impossible: Sayings, for the Most Part in Parable, from the Letters and Journals of Lilias Trotter of Algiers*. (New York: Macmillan, 1938), 65–66.

Lilias Trotter, *Mountain Scene*, watercolor, date unknown.

Gerard Manley Hopkins

*Gerard Manley Hopkins was an English poet who, at age twenty-two,
became a Roman Catholic, and later, a Jesuit. He died of typhoid
fever in 1889 at age forty-four.*

WHY DID GOD CREATE? – Not for sport, not for nothing. Every sensible man has a purpose in all he does, every workman has a use for every object he makes. Much more has God a purpose, an end, a meaning in his work. He meant the world to give him praise, reverence, and service; to give him glory. It is like a garden, a field he sows: what should it bear him? Praise, reverence, and service; it should yield him glory. It is an estate he farms: what should it bring him in? Praise, reverence, and service; it should repay him glory. It is a lease-hold he lets out: what should its rent be? Praise, reverence, and service; its rent is his glory. It is a bird he teaches to sing, a pipe, a harp he plays on: what should it sing to him? etc. It is a glass he looks in: what should it show him? With praise, reverence, and service it should show him his own glory. It is a book he has written, of the riches of his knowledge, teaching endless truths, full lessons of wisdom, a poem of beauty: what is it about? His praise, the reverence due to him, the way to serve him; it tells him of his glory. It is a censer fuming: what is the sweet incense? His praise, his reverence, his service; it rises to his glory. It is an altar and a victim on it lying in his sight: why is it offered? To his praise, honor, and service: it is a sacrifice to his glory.

The creation does praise God, does reflect honor on him, is of service to him, and yet the praises fall short; the honor is like none, less than a buttercup to a king; the service is of no service to him. In other words he does not need it. He has infinite glory without it and what is infinite can be made no bigger. Nevertheless he takes it: he wishes it, asks it, he commands it, he enforces it, he gets it.

Gerard Manley Hopkins, ed. Christopher Devlin, *The Sermons and Devotional Writings of Gerard Manley Hopkins*. (United Kingdom: Oxford University Press, 1959), 238–239.

Lilias Trotter, *Leaves of Gold*, watercolor, 1906.

Augustine of Hippo

*Saint Augustine (354–430) was bishop of Hippo
and one of the Latin Fathers of the Church.*

HOW CAN I TELL YOU of the rest of creation with all of its beauty and utility, which the divine greatness has given to man to please his eyes and serve his purposes? . . .

Shall I speak of the manifold and various loveliness of sky and earth and sea; of the plentiful supply and wonderful qualities of light, of sun, moon, and stars; of the shade of trees; of the colors and perfume and song; of the variety of animals, of which the smallest in size are often the most wonderful – the works of ants and bees astonishing us more than the huge bodies of whales?

Shall I speak of the sea, which itself is so grand a spectacle, when it arrays itself as it were in vestures of various colors, now running through every shade of green, and again becoming purple or blue? Is it not delightful to look at the storm and experience the soothing complacency which it inspires by suggesting that we ourselves are not tossed and shipwrecked? What shall I say of the numberless foods to alleviate hunger, the variety of seasonings to stimulate the appetite which are scattered everywhere by nature, and for which we are not indebted to the art of cookery? How many natural herbs are there for preserving and restoring health? How graceful is the alteration of day and night! How pleasant the breezes that cool the air! How abundant the supply of clothing furnished us by plants, trees, and animals! Can we enumerate all the blessings which we enjoy?

Augustine of Hippo, *City of God*, in *Nicene and Post-Nicene Fathers*, Vol. II. Ed. Philip Schaff. (NY: Charles Scribner's Sons, 1907), 504.

Lilias Trotter, *Desultory Bee*, watercolor, 1907.

George MacDonald

George MacDonald was born in 1824 in Aberdeenshire, Scotland.
He left the ministry to pursue a literary career and wrote over
fifty books to support his family of eleven children.

EVERY FACT IN NATURE is a revelation of God, is there such as it is because God is such as he is; and I suspect that all its facts impress us so that we learn God unconsciously. True, we cannot think of any one fact thus, except as we find the soul of it – its fact of God; but from the moment when first we come into contact with the world, it is to us a revelation of God, his things seen, by which we come to know the things unseen. . . .

What idea could we have of God without the sky? The truth of the sky is what it makes us feel of the God that sent it out to our eyes. . . . In its discovered laws, light seems to me to be such because God is such. Its so-called laws are the waving of his garments, waving so because he is thinking and loving and walking inside them. We are here in a region far above that commonly claimed for science, open only to the heart of the child and the childlike man and woman. . . . Facts and laws are but a means to an end; in the perfected end we find the intent, and there God. For that reason, human science cannot discover God; for science is but the backward undoing of the tapestry-web of God's science; it will never find the face of God, while those who would reach his heart will find also the spring-head of his science. The work of science is a following back of his footsteps, too often without appreciation of the result for which the feet took those steps. If a man could find out why God worked so, then he would be discovering God; but even then he would not be discovering the best and deepest of God; for his means cannot be so great as his ends. ➤

Adapted from George MacDonald, *Unspoken Sermons*, Series III. (Urbana, Illinois: Project Gutenberg).

Lilias Trotter, *Blossom in the Desert*, watercolor, ca. 1896.

RHYS LAVERTY

Breakwater

In the Channel Islands, a crumbling
jetty protects a way of life.

but not of the United Kingdom. It falls within the Bailiwick of Guernsey, which, alongside Jersey and the Isle of Man, is one of the United Kingdom's Crown Dependencies. Despite being British, it lies just ten miles from the French coast, compared to sixty from the English. At just three and a half miles by a mile and a half in size, with a population of only around two thousand, "The Rock" is a true haven – a small-scale place in a large-scale world. While the world suffered from months of lockdowns during the Covid-19 pandemic, life on Alderney continued (mostly) as normal.

Life seems slower here. The shops close on Sundays. And at lunchtime. And at whatever other time takes the shopkeeper's fancy. Air travel is infrequent, and a sudden fog rising off the Atlantic, or a gremlin in the engine of one of the only two commercial planes that fly there, is enough to keep you stranded for a day or two. There is only one cash machine. It might strike you as the last holdout of a more traditional way of life. But its official websites boast of its appeal to remote workers, its superfast broadband and great transport connections. You might want to get away from the world in Alderney, but the world will find you. Alderney's a proud part of the large-scale world, and it has been since that fateful decision in 1847.

When the planes are working and the air is clear, the first thing you see when you fly toward Alderney from the north is the breakwater.

Today, the breakwater is a half-mile-long wall of stone stretching out into the English Channel. It provides shelter to the island harbor, which in season is full of sailors and yachtsmen. Many of them arrive on Alderney's shores to test themselves on its tides, which have earned a reputation as some of the most challenging in the world. It also shields Braye, the island's main beach – a serene

I N 1847, A DECISION was made in Whitehall. It would change the destiny of the small island of Alderney, 150 miles away.

Today, as back then, no one knows about Alderney. From what I can tell, the residents like it that way – mostly. The only reason I know the place is my ancestry. On Dad's side, my family has been there since at least the 1500s. It's the northernmost of the Channel Islands – part of the British Isles,

Rhys Laverty works as senior editor at Ad Fontes *and managing editor at the Davenant Press. His writing has appeared in* The Critic, Mere Orthodoxy, *and* World. *He writes a regular Substack,* The New Albion. *He lives in Chessington, United Kingdom, with his wife and three children.*

Neil Howard, *Alderney Breakwater*, 2016.

white crescent of sand, shimmering like the sharp edge of the sea. Waves will splash over the top of the breakwater most days; in stormy weather, walls of water plow into it, hurling foam dozens of feet into the air. On clear days, though, anglers dot the breakwater's top, and tourists can be found walking the length of it, peering over the edge into the chop and churn of the ocean below.

On those days, between May and October, there's another sight you're likely to see on the breakwater: a large red mobile crane. At the end of its cable hangs a cage, and into the cage go the divers and engineers. "Some of the seaward face can only be viewed from the sea," Marco Tersigni, one of the infrastructure officers for the States of Guernsey, told me. "Underwater inspection is carried out by a diving contractor during the summer and in the shoulder months. The crane and the cage are used for lowering the divers down to the water. On occasion they can be used to suspend contractors on the seaward face of the breakwater, normally to repoint open joints in the masonry." I can't help but think of miners descending pit shafts when picturing these men

being lowered down the breakwater's north side to assess the endless assaults of the waves.

When Whitehall signed off on this edifice in 1847, it was to be the stony backbone of a new "harbor of refuge" on Alderney – a euphemism for "naval base" that fooled precisely no one. Despite its tiny size and population of barely more than a thousand, Alderney's uneasy proximity to France made it strategically significant. Its natural fortification by violent tides also helped. Over thirty years after the Napoleonic Wars, Anglo-French relations remained tense. Ten miles over the water at Cherbourg, the French were clearly arming themselves. And so, far away from its inhabitants, at the heart of the British military establishment, the decision was made: Alderney would become a "Gibraltar of the Channel" – a grand ambition. A few decades later, a critical French writer would say this was a name "which could only have been born at the bottom of a bottle of sherry." He would be proven right. The bombastic, bellicose endeavor would, in many ways, be the death of the old Alderney. And yet, nearly two centuries later, its legacy is a lifeline for its modern inhabitants.

Neil Howard, *Braye Bay, Alderney*, 2018.

ON FEBRUARY 12, 1847, a ceremony involving cannon fire, flags, prayers, a band, and musketry was held at Grosnez Point, the northern outcrop which would anchor the structure. The foundation stone, a huge chunk of island granite, was dropped thunderously down a cliff. That night, all over the island, people celebrated with fireworks and feasting. Work began in earnest the next day.

The first three years were spent dumping piles of granite into the consuming waters of the Channel to form the base. In places, the sea was up to 150 meters deep. A new inner harbor had to be constructed, along with a railway to access the island quarry and transport stone. Work on the breakwater proper could not commence until 1850.

The original designs for the harbor are truly astounding. The half-mile that remains of it today may seem formidable, but the Admiralty first envisioned a length of nearly two miles curving around the northwest of the bay; this would partner with a shorter breakwater of half a mile or so reaching out from the northeast. Together, they would cast a protective ring of rock around the imagined harbor, resembling a mother's arms or a crab claw. Here, imagined the men in Whitehall, a mighty fleet could be housed and British merchant ships could seek refuge from French privateers.

The breakwater was to be accompanied by mighty fortifications as well. In the decade or so after the breakwater's construction began, about a dozen military forts were built on the island, along with other battery and barrack buildings. The massive influx of workmen and soldiers required for all this trebled the island's population from 1,038 in 1841 to 3,333 in 1851; in 1861, it peaked at 4,932.

Plunging stone and brick into the English Channel in order to build a breakwater proved challenging. For one thing, the engineering firm in charge, Jackson and Bean, had no expertise in breakwaters, only canals and railways. One writer described the process some years later:

The storms of winter tore down the work of the previous summer; the gales of the summer broke down the staging and machinery employed in the construction and swept all into the sea. Still the contractors held on, fighting against all these difficulties with untiring determination, learning lessons for the future by the disasters of the day.

And yet, against all this, the engineers succeeded – somewhat, at least. The breakwater is still standing. In 2018, it was selected by the Institute of Civil Engineers as one of two hundred structures worldwide that have shaped the world and transformed lives. Yet the original plans were never realized. The second northeast breakwater never even began. The main breakwater, at its peak in 1864, reached 1,471 meters – almost a mile. It had cost, in today's money, more than 177 million dollars. Yet the structure became, in the worst way, more like a work of art than a civil engineering project – abandoned rather than completed. Within a few years around a third of it, impossible to maintain, had been given up to the sea, bringing the structure to its current length. Ironically, after being intended to protect British ships from the French, the abandoned stretch of breakwater formed an artificial reef that made entry to the harbor hazardous for many vessels, often pushing them to land at Cherbourg instead.

In the end, the breakwater, the forts, all of it, were for a war that never came to pass. In 1853, France became an ally in Crimea; in 1871, it suffered an unprecedented defeat in the Franco-Prussian War, and was crippled by debt and German occupation. What's more, developments in artillery meant that, by the time many of Alderney's forts were finished, they were already redundant – white elephants all. The "Gibraltar of the Channel" it was not.

All of this came at the cost of the island's centuries-old culture and way of life. Dramatic changes faced a community in which everyone, as English observers condescendingly recorded, was

a small landowner and lived "almost exclusively on the produce of their soil." What else could happen when you quintuple an island's population in just a few decades and reshape its coastline by artifice? This process was already underway in Alderney prior to the fortifications, but the breakwater accelerated it. A garrison of some four hundred soldiers was stationed there between 1800 and 1815. During this time, as various travel writers record it, the use of the French language in general, along with the island's native Norman French patois, began to decline. Still, when Queen Victoria visited in 1854 to inspect the building work, banners in the street read *Dieu sauve la Reine.* But this was the swan song of Alderney's island culture. By 1901, one Channel Island guide could state confidently that "Alderney is the least

French of all the Channel Islands. The local patois, which differs considerably from the vernacular of the other islands, is rarely heard nowadays."

THE LOCALS – the Aurignais, to give them their proper name – had never held much stock in the government's grandiose military plans, but even they did not object to the changes these plans brought about. Prior to the breakwater, Alderney's economic peak seems to have been the late 1700s and early 1800s, when smuggling and privateering abounded, and "the little Alderney privateer reaped a good harvest in a bad way." But by the mid-1800s, the rise of state navies brought privateering to a close and the island was all the worse for it – its exports being chiefly potatoes and lobster. The economic benefits of a

Alderney Harbor, plan of breakwater and Admiralty property, 1878.

large army garrison and constant construction work were not sniffed at by many. Alderney's story was one repeated the world over in this era: a central government made a great modernizing decree, giving little thought to how it might upend the traditional culture of the locals, but found that the locals, despite their reservations, were nonetheless happy to acquiesce to the prospect of wealth and comfort.

All this seems so long ago – a different world. The redundant forts have either become lavish residential properties, are used for storage, or have simply been left to the elements for adventurous youngsters to explore. Victorian Whitehall's choices may seem consigned to being historical curios, a reminder of the buffoonery of government officials. But they are also the visible marks of the many ways life in Alderney was transformed. After the breakwater was built, the mail came twice a week. It opened the doors to a much wider import trade than Alderney had ever experienced before.

And then there is the breakwater – a King Cnut–like structure, testament to madcap military ambition, and somehow essential to the life of this small island. Did anyone in the Admiralty consider how the breakwater would be maintained two hundred years hence? I doubt it – but it is a reality that Alderney must live with.

EVERY YEAR, between November and April, the breakwater is given up to the sea; then, between May and October, the damage must be swiftly assessed and repaired. "Depending on the type, scale, and location of the repair, we normally use concrete either with or without reinforcement," Marco Tersigni told me. "For repair of underwater joints carried out by the diving contractors we use marine mortars, and normal mortars for above-water joints." The patchwork nature of the seawall, the changes in the tone and shape of the masonry, are visible as you walk along it, and perhaps even more so from a distance. Since 1987, when the Bailiwick of Guernsey took over ownership of the breakwater from the United Kingdom, more than 31 million dollars have been spent on its upkeep. Much may have changed since 1847, but the Atlantic Ocean has not. Yet the funding of the breakwater – like the funding of much on Alderney – is a continual bone of contention between the island's own administration and that of the wider Bailiwick. Pouring stone into the sea means pouring money in too.

What would life be like without the breakwater? The ferocious tides mean there is no sufficient natural harbor. Longis Bay, on the island's southern, French-facing side, once served as a harbor, and was supposedly the site of the island's earliest settlement. However, it is just as exposed as Braye. Legend has it that the ancient settlement there was wiped out by some kind of natural disaster, interpreted by the locals as divine judgment.

A breakwater is a necessity for a properly functioning harbor on Alderney. And, in 2024, a properly functioning harbor is essential for life on the island. Being a small-scale place in a large-scale world is well and good, but that world can't be avoided entirely. Without the breakwater, towering waves would dash the *Trinity*, the island's twice-weekly supply ship, to pieces on the rocks. The yachtsmen ("boaties," to the locals) and beachgoers crucial for the island's tourism industry would cease to find Braye an idyllic holiday destination, and perhaps go elsewhere altogether.

Today, then, the Aurignais have no choice: the breakwater, bequeathed to them by the hubris of some Victorian admirals (and their bottle of sherry) in a decision made far away and long ago, has become indispensable. It must be repaired and repaired again, ad infinitum. ⇝

Are You a Tree?

Or are you a potted plant?

JOY MARIE CLARKSON

I HAD MOVED HOUSE AT LEAST once a year for seven years straight. It is simply the way of life during higher education, the path I chose in my early twenties. When the short years of an undergraduate degree expire, one is sent into a seemingly endless game of musical chairs; if you're not moving for a new degree or a new short-term job, you're moving to find a cheaper place to live or a better roommate, or simply bending yourself to the will of campus housing. It became wearying, but as the years wore on, I began to strategize. In preparation for my move to each new domicile, I kept a few prized possessions, pictures, sentimental things, and valuable household items to be loaded into a single cardboard box. I'd collected these objects in

---

Joy Clarkson holds a PhD in theology from the Institute for Theology and the Arts at the University of Saint Andrews. She hosts Speaking with Joy, *a popular podcast about art, theology, and culture, and writes books, including her most recent,* You Are a Tree, *from which this excerpt is taken.*

William Morris, fabric designs,
Tree Frieze, top, *Tulip Frieze,* bottom, ca. 1876.

hopes that one day I'd have my own home, where they could be of use or gather dust on a decorative shelf. "Have nothing in your house which you do not know to be useful, or believe to be beautiful" wrote William Morris, and I tried to follow his maxim. But each year as another June rolled around, a less idealistic proverb formed itself in my mind: have nothing in your apartment which you do not know to be disposable, or believe to be easily transported.

For me, the necessity of portability did not begin in college. If you were out to dinner with my mother and asked her where our family was from, she would grin and with a twinkle in her eye recite the (to me) familiar formula: "We've moved sixteen times, six times internationally" to the consternation of the listening party. Each of my siblings was born in a different state or country, and until my parents moved into our family home in Colorado, they had never stayed in one house for more than three years. The possibility that next year, next month even, I might need to move again has always been more than that to me; it is a probability, no, an inevitability. And so, when, during my studies, I settled into a flat for more than a year – twenty-seven months to be exact – it felt almost miraculous. But eventually my studies came to a close, and it was time to move again.

I sat on the stoop of my flat that breezy September morning, a shambolic mess of half-packed cardboard boxes within, and sighed. I was sad in a tired way. I had grown to like the overgrown garden our flat shared with the letting agency below: the pear trees that were supposedly grafted from medieval trees, the overburdened trellis of roses with limbs arched in blossomed exhaustion, the tree that produced six perfect red apples each fall (no more, sometimes less), even the tropical tree with large, waxy leaves that seemed not quite at home in this gusty Scottish climate. I looked on them with a mixed pleasure. I envied the impassive stability of these trees; they would go on sprouting, blossoming, changing colors with the seasons, not caring whether I stayed or left, lived or died. Oh, to be so stubborn in one's being and one's needs, so sure of one's literal place in the world.

I am a potted plant, I said to myself. *Always ready to be moved, never mingling my roots with those of my neighbors, a stranger to solid ground.*

This thought fell into my mind like a blunt object. Inside my house was a small potted plant that I had taken pains to keep alive during the final throes of completing my thesis, like a talisman of my own survival. I had been contemplating what to do with it when I left and considering throwing it away. It had taken on a wild, stringy appearance no matter how I groomed and clipped it, as though in protest against its modest pot, signaling to me that it wanted to get out and spread itself into welcoming soil. The metaphor continued to unravel itself within my knotted stomach. *Perhaps I am a plant that has grown too large for its pot, a plant that if it does not find real soil to set its roots in soon will become awkward and sad, limbs reaching pleadingly toward the sun at the window, wanting to feel the worms and wetness of early morning, but always kept outside of such experiences.* Lately, the sight of the plant had begun to inspire a plaintive despair in me; I had tried to plant indoor plants outside before, but they soon died, their roots shocked by the experience. Darkly, I thought to myself that perhaps I was the same. Perhaps after all these years of life with portable roots, I was no longer capable of natural rootedness. Perhaps if you tried to plant me in a particular place, I would shrivel up and die, not ready for the exposure of pure obligation to a place. I longed for a place to belong, to be entangled with, but felt in my bones incapable of such a thing.

I am not the only potted plant. In centuries past, the odds were good that you would grow up, marry, and work within a short distance of the same place where you were born. Now it is less likely to be the case. Writing in response to the moral and economic desolations of World

War II, Simone Weil drew on this metaphor in her book *The Need for Roots*, describing the modern condition as one of rootlessness, a lack of meaningful community, work, and belonging, a loss many of us feel in our souls. Some have tried to react to this sense of itinerancy constructively, by choosing a place to put down their roots for good. The notion is noble but comes with its own angst. It is very difficult to belong to a place, to not be able to escape it, to be bound to petty church politics, racist neighbors, the limitations of *this place*. And how does one choose a place? There is a loneliness of knowing that your rootedness is a chosen rootedness, not the inheritance of love and history. This was a pain I first put my finger on in

the golden idealism that first comes with reading a Wendell Berry novel. After a period of wistful desire to take up farming and only use a typewriter, I began to feel a sassier question rising to my pen: it's all fine and good, Mr. Berry, but what if I have no ancestral farm? And, after all, how far back do I have to look to discover the farm isn't so ancestral after all?

The feeling of rootlessness stretches much farther back past our present predicament. One credible description of history is a long legacy of displacement; of winning and losing land, of conquering and being driven out, of building homes and having them destroyed, by war or time, greed or boredom. Rootlessness is not merely a feature of the modern condition but also of the human condition.

I felt this keenly when I first read Saint Augustine's *Confessions*, where he touches this ancient wound in a surprisingly vivid way. The North African saint whose words and ideas have echoed down through the centuries described human nature as being characterized by a kind of restlessness. He famously writes in his *Confessions*, "We are restless until we find our rest in thee."

Augustine was what we might call a third-culture kid – the son of a Christian North African mother and a pagan Roman father – never quite fitting anywhere. Reading the story of his early life in *Confessions* is strikingly relatable to us sufferers of (post)modern malaise. As a young man he reinvented himself again and again. First, he fashioned himself as a hedonist and a social climber, intoxicated by romance and every pleasure that came his way. Made a bit sick by his own overindulgence, Augustine turned to a restrictive lifestyle, joining a gnostic cult with strict rules for living and high-minded ideas about the spiritual world. Finally, and perhaps most tragically, he fell in love, taking a lover with whom he had a child, and whom, by all accounts, he never gave up loving. In each act of his recounted life, there is a tragic sense of longing, unsatiated desire. When I read his

William Morris, design for a printed textile, 1883.

fraught words, I can't help but feel that Augustine, too, was a potted plant, withering with desire.

But Augustine took a different metaphor as the interpretive key for his life: a journey, or, rather, an exile. Sarah Stewart-Kroeker writes "Augustine's dominant image for the human life is *peregrinatio*, which signifies at once a journey to the homeland (a 'pilgrimage') and the condition of exile from the homeland." All of life for Augustine was shaped both by this search for the homeland and the feeling of exile; he was a potted plant searching for welcoming soil. This feeling characterizes not only the ethos of his theology, but also the arc of his own personal narrative. In Augustine's story, I found resonances of my own: the desire for rest and rootedness mixed with the sense of exile and strain toward a place of belonging. Here, I began to find myself mixing metaphors. I am a potted plant; I am a pilgrim. The image it presented to me was awkward and funny, like Tolkien's glacially slow and meandering tree-people, the ents. What could flourishing look like for this mixed metaphorical life? How can one succeed as both a pilgrim and a tree? Of a promising person we say *they are going places*. We do not say that of a successful tree. A successful tree stays put. It has roots. It bears fruit.

Somewhere along the way, I discovered that this mixed metaphor is at the heart of one of the Bible's most famous passages: Psalm 1. This is what it says:

> Blessed is the one
> who does not walk in step with the wicked
> or stand in the way that sinners take
> or sit in the company of mockers,
> but whose delight is in the law of the Lord,
> and who meditates on his law day and night.
> That person is like a tree planted by streams
>   of water,
> which yields its fruit in season
> and whose leaf does not wither—
> whatever they do prospers.

> Not so the wicked!
> They are like chaff
> that the wind blows away.

> Therefore the wicked will not stand
>   in the judgment,
> nor sinners in the assembly of the righteous.

> For the Lord watches over the way of the righteous, but the way of the wicked leads to destruction.

The blessed person walks, like a pilgrim (v. 1), but the blessed person is also like a tree (v. 3). At first the psalm begins as a simile, but it unfolds the likeness in metaphor; the righteous are not only like the tree, they *are* planted, yielding, prospering. At the heart of these two images is not only the (not) nature of metaphor but also some of the central tensions of what it is to be a human. We flourish in rootedness and fruitfulness, but that rootedness is always temporary, interrupted by death. And even in life we are driven by longings this world never seems capable of satisfying. By reflecting on the properties of trees and journeys, and carrying them over to the human condition, we might discover new ways of understanding ourselves. And even in the ruptures of the metaphors, those places where there is not correspondence, we might discover and articulate those ruptures and noncorrespondences in the human experience that cause us most discomfort and pain. In speaking about them, in giving them the form of images in our mind, we might find ourselves consoled, or drawn onward. The seemingly contradictory images of trees and journeys invite us to consider what it is like to be human, to flourish, to live well in the contradictions of human nature, with the desire for eternity in the confines of mortality, roots in the ground and branches arching their weary arms toward their heavenly home. ⤳

Joy Marie Clarkson, *You Are a Tree: And Other Metaphors to Nourish Life, Thought, and Prayer* (Bethany House Publishers, 2024), 13–16, 30–34. Used by permission.

The Leper of Abercuawg

In a thousand-year-old Welsh poem, an outcast seeks comfort in the wild.

DAVID MCBRIDE

Malcolm Edwards, *Mignient*, watercolor, 2020.

I N 1976, the poet and Anglican priest R. S. Thomas was asked to give a lecture at the National Eisteddfod, the Welsh annual festival of art and literature. Thomas, one of the finest twentieth-century poets, was a man of many paradoxes: though an ardent nationalist and advocate of the Welsh language, he did not learn to speak it until his early thirties and wrote poetry only in English; though he ministered as a priest for forty years, he struggled with faith for even longer, eventually describing himself as a "retired Christian." His lecture asked, "*Lle mae Abercuawg*? – Where is Abercuawg?"

Abercuawg is the unidentified setting of the ninth-century poem "Claf Abercuawg," told from the perspective of a *claf,* a sick man – most likely a leper – who finds himself exiled to a desolate wood because of the stigma of his disease. With his old life gone and his body failing him, the speaker meditates on those things that endure: sin, redemption, and the mysteries of God. A thousand years after its composition, "Claf Abercuawg" continues to fascinate readers. The unknown author's opaque, allusive style prefigures twentieth-century modernism; the experience of the leper reflecting on the pristine natural world he encounters attracts readers disenchanted with an urban, mechanized modernity.

For Thomas, Abercuawg represents the world as it should be, a Wales lost which can only be regained by triumphing over contemporary indifference, the machine, and the English language. *Lle mae Abercuawg? . . . ni welaf ystyr i'm bywyd, os nad oes y fath le ag Abercuawg, tref neu bentref y mae'r cogau'n canu ynddo.* "Where is Abercuawg?" he asks. "I do not see meaning in my life, if there is not a place such as Abercuawg, if there is no stead or village where the cuckoos sing." For the leper of Abercuawg, the pain of separation and the anxieties of self-reflection alone in creation, where the only thinking beings are him and God, is resolved in a renewed faith: a recommitment, in spite of everything, of his life to Christ. The author of "Claf Abercuawg" positions himself both within an existing tradition of Welsh Christian nature poetry, and outside it. Such poems tend to be allusive, with the poet taking on the character of the speaker, and offering bits of wisdom or gnomic sayings about the human and natural worlds, interspersed with straightforward descriptions of the landscape. Part of the art of these poems is transforming the particular into the universal. In one, a grieving young woman, having lost her family and home in war, exclaims *dygystud deurud dagreu,* "tears wear away the cheeks," connecting the character's own experience of loss to what was certainly a common occurrence in early medieval Wales. Such gnomes (from the Greek for aphorism) are not unique to early Welsh poetry; they are found in Old English works like "The Wanderer" or "The Seafarer," as well. In most Welsh saga poetry, the speaking character has a backstory the audience would know; this does not seem to be the case with the wise leper in "Claf Abercuawg."

Our exiled companion is a nobody, his fame and status all in the past. He laments to his listeners – the poem would have been recited – that he "cannot lead a troop." That he misses war might strike the modern reader as odd, but his audience would have immediately understood such sentiments. Ninth-century Wales, for all its literary merits, was a violent place, even by medieval standards; one could make a rough parallel to the martial culture of its near-contemporary *Beowulf* or the *Iliad.* A nobleman was expected to provide for his warband, which in turn protected his interests and property. By the time "Claf Abercuawg" was composed, such a state of affairs had abated in England, but survived in Wales much longer due to widespread cattle-stealing – that time-honored way of transferring wealth and power – and the lack of centralized authority that would prevent such endemic violence. In stressing his lost status as a nobleman, from the forefront of this martial culture, the leper makes clear how far he has fallen.

He has indeed fallen far enough that, without position or human community, he lacks even a name. Lepers were often exiled from early societies for fear of contagion – a stigma reinforced by the common belief that leprosy was a divine punishment for some unknown transgression. The doubled exile – physical and spiritual – was total. One liturgical document, a few centuries on from "Claf Abercuawg," describes a ritual separating the leper from society. The priest declares: "I forbid you ever to enter churches, or to go into a market, or a mill, or a bakehouse, or into any assemblies of people"; the leper is forbidden to dine with non-lepers or to speak to people in public, even when addressed, except from a distance. He is not cut off just from his old community, but from human society itself. It is unsurprising, then, that many lepers made their homes in places remote from human habitation. Even there, our leper's condition haunts him. We first encounter him at the base of a hill: his soul "longs to sit for a while" on its summit, but his body is unable to take him there, even though it is not far. The pleasures of summer, the breeze, the colors of the flowers and the wood, bring only

David McBride is a third-year PhD student in the Department of Welsh and Celtic Studies at Aberystwyth University. His interests include Celtic linguistiscs and the interplay of early Medieval Insular literature and history.

passing pleasure to him, as he admits he runs a fever. His home, he tells us, has fallen into disrepair or disuse.

Reminiscing on his exile during a lull in his afflic-tion, he is interrupted by a cuckoo's call – summer is a-coming in. He offers his first gnomic observation: "He that giveth too much is better than he who is a miser." Even as he begins to reflect on the flaws of his past life, he finds a brief release in the beauties of nature: "Chattering cuckoo,

may it sing forever!" Some of the earliest Welsh poetry, a metrical lyric known as the "Juvencus Nine," dwells on how the grass and trees, if they sang in chorus, would not be sufficient to recount all of God's glories. This perhaps calls upon imagery from Psalm 96: "Sing a new song to the Lord, all the earth." We find similar expressions in Augustine: "A lizard catching flies, or a spider entangling them as they rush into her nets, oftentimes arrests me. . . . From them I proceed to praise you, the wonderful Creator and Disposer of all things." Our leper's thoughts do not yet rest explicitly on the Creator, but he too finds joy in creation.

Suddenly, the pleasures of beholding the natural world turn to misery. The leper is keenly aware of how much he has lost. The cuckoo flitters about, giddy, searching for a mate, and the leper is immobile, alone, once again in pain. As the bird longs for another, the speaker longs for everything he once had. The bird can look forward to finding a companion, but what hope does the leper have of restoring the life he has left behind? The air is cold, the night lengthens. His heart chafes at his illness, and he longs for death "from his disease

and age." He contemplates the shore, marveling at the sun's reflection on the flow of water, though this does not last long either, as he is interrupted by the sounds of birds being chased by baying hounds and laments that his illness does not allow him to go to war. Perhaps he has heard his own former self in the sounds of the hounds and the hunters chasing their prey.

The sun breaking through the clouds – a rare occurrence in the region, even in summer – brings the leper's gaze back to the hill which his soul longs to ascend. But this light brings no comfort. He declares "my fever has chosen me," perhaps indicting his own past actions for the illness he now suffers. It is here that his gnomes are adduced to criticize human behavior, perhaps with one eye to himself. Contemporary readers might find alien the belief that God punishes people on earth with bodily disease for their sins. Others of his atti-tudes are easier to make sense of: today as in the ninth century, being forced to face one's mortality with a deadly or chronic illness does give a new perspective to life. It is not clear that the leper views his disease as a punishment. He muses: "The idle are wont to arrogance"; "The wise man desireth not discord"; "Patience encompasseth understanding." Reading between the lines, we glimpse a condemnation of the noble lifestyle the leper strove to emulate in the past.

This critique of the heroic ideal is not unique in early Welsh poetry. In the slightly earlier saga poem scholars call "Gwên and Llywarch," an aged father, Llywarch, convinces his last surviving son, Gwên, to defend a border-ford against a greater force. Llywarch boasts of his own youthful exploits to convince his son, who might otherwise have been dissuaded since all his brothers have previously died in battle and the odds are against him. Gwên sees through his father's bravado, but, duty-bound to obey and hoping to gain glory, goes to battle nonetheless. He is slain. Llywarch elegizes him, attacking his own speech for leading to his son's untimely demise, but does not extend his condemnation to the heroic ideal itself. Even awash in pain, he commends his son's bravery: he has died but did not retreat. There is no such contradiction at work in the leper's critique of his society. His world has left him behind, and he has left its standards behind in turn. Gwên, in youth and the pride of his father, could gain glory or victory so long as he was brave, but the leper has nothing to lose. He has no treasures to store up on earth, no way to gain earthly fame or possessions.

Now the leper finds himself in winter, and with the distractions of the surrounding landscape diminished, he focuses on more spiritual questions. "The heart is deceitful above all things. It is full of deceit, and wicked works. There shall be grief when it is cleansed." Deceit is no small part of how the leper would have made his living before his illness: thieving and scheming played an indispensable role in the cloak-and-dagger politics of medieval Wales. The significance of literal daggers is hard to overstate: punishments such as blinding and castration were not uncommon. We might imagine the leper reflecting on his own past sins, then, when he denounces the lies and immorality of power and fame. "When the Lord judges on

the long day the false will be in gloom, the true in light." He seems to reject his former life and the earthly glories he had wished to attain: "The attacker is ragged." The noble warrior is now in rags slipping off his weakened body. All his efforts have led to this.

"My heart rubs raw from depression tonight. . . . The cheek cannot hide the heart's grief." Would these spiritual realizations have occurred to him without his exile and affliction? He has an acute sense of his own unworthiness: "God allows no good for the hapless man." He even feels he must be "hated by God above," who gives him "only sorrow and care." He has tried, it seems, the usual treatment for his contagion – going to a monastery to partake in a "lepers' mass." The purpose of these services was not to cure the lepers, however, but to console them against hopelessness, drawing on the words of the apostle Paul: "For I reckon that the sufferings of this present time are not worthy to be compared with the glory which shall be revealed in us."

"Claf Abercuawg" ends abruptly, in words that suggest the speaker has internalized the suffering Saint Paul speaks of. "May God be kind to the outcast," the leper says, in what is both a plea for grace and an acceptance of his condition. This is, conforming to the tropes of gnomic poetry, a particular experience with universal implications. The earliest word for leprosy recorded in Welsh is *gwahanaint*, literally, "the disease of separation." Given this reading of the poem, the word takes on a new meaning: Christians believe we are all separated from God through our own sinfulness, an illness as total and as painful as the leper's. We long to climb the summit of a bright hill, to find unity with God in heaven, but we find our own weakness continually works against the attempt. By our sin, and our consequent affliction, we have created a gulf between us and God that we cannot cross. But he can. With God's grace, all things are possible, and even the shame and solitude of exile can appear as a gift. Even as the leper in this poem

claims to be hated by God, he knows this is a lie. Outcast from the human world he once knew, alienated from the natural one, the leper finds himself wishing to be drawn near to God.

The "sickness of separation" that "Claf Abercuawg" wrestles is still with us: many people whose lives are mediated by technology and deprived of meaningful community will relate to the isolation of the leper. It was this loneliness, fostered by the encroachments of the "machine" and the spiritual desolation of a secularized, scientific worldview, that formed much of R. S. Thomas's poetry. The priest-poet spent his life seeking "the far side of the cross"; trying to find a path to faith through nature and introspection in the face of God's apparent absence. In one of

his late poems, "The Flower," Thomas speaks of the "unseen flower" of faith – a faith he arrived at, as the leper did, through surrender to God even in his apparent silence. "I gave my eyes and my ears," Thomas writes, "and dwelt in a soundless darkness in the shadow of your regard." When we cannot see God in our social surroundings, or the natural world, or in our own fallen condition, it is in our "hapless" nature that God comes to save us. When the leper of Abercuawg writes that "God will not undo that which He does," it is both an expression of despair in his own condition and a declaration of ultimate hope. God's promise of redemption is one on which we can, and must, rely. Until then, as Thomas once wrote, "the meaning is in the waiting."

Malcolm Edwards, *Braich y Waun*, watercolor, 2020.

Claf Abercuawg
Translated by David McBride

My soul longs to sit for a while on this hill
Alas, I am unable to take one step.
My journey is not long, my home is deserted.
The wind pierces, the cow-paths are bare—in summer,
when the trees obtain their comely color.
I shiver with fever today.
I am not agile, I cannot lead a troop,
I cannot move about.
While the cuckoo is happy, let it sing!
A chirpy cuckoo hails the dawn
Generous its words in the fields of Cuawg.
He that giveth too much is better than he who is a miser.
In Abercuawg, cuckoos sing
on branches bedecked with blooms.
Chattering cuckoo, may it sing forever!
In Abercuawg, cuckoos sing
on branches bedecked with blooms.
Woe to the leper, hearing them always.
In Abercuawg, cuckoos sing.
It brings my heart anguish
that I have no more a friend to hear them with me.
I have listened to a cuckoo on an ivy-covered tree.
My clothing has become looser.
Grief for what I loved is greater.
On the peak above the rustling oak
I listened to the birds' call.
Tuneful cuckoo, all remember what they love.
Always crooning its carol, its cry full of yearning,
About to flit, of flight like a hawk
Is the eloquent cuckoo in Abercuawg.
Vocal the birds; wet the valleys.
The moon gives light; the night is frozen in stillness.
My heart chafes, my disease grieves me.
Vocal the birds; wet the valleys; this night is long.
A good thing should be held onto.
I am owed my sleep from my disease and age.

Malcolm Edwards, *Blaen y Nant*, watercolor, 2020.

Vocal the birds; wet the roof-tile.
Leaves fall; the exile is dispirited.
I will not deny I am ill tonight.
Vocal the birds; wet the strand.
Bright the sky; broad the wave.
My heart withers from longing.
Vocal the birds; wet the strand.
Bright the wave of broad movement.
That which was loved in my youth—
I would love if I were to have it once more.
Vocal the birds in the hills of Edrywy.
Loud the hounds' baying in the wastes.
Vocal the birds again.
Early summer—each growth is fair.
When warriors rush to battle,
I do not go; my illness forbids it.
Early summer—fair the borderland.
When warriors hasten to the battlefield
I do not go; my affliction burns me.
Hazy the hilltop in the sun; broken the ash's branches.
From the estuaries a bright wave flows away.
Mirth is remote from my heart.
Today for me is the end of another month
In the leper-house where I am abandoned.
My heart chafes, my fever has chosen me.
Clear is the sight of the water.
The idle are wont to arrogance.
My heart chafes; sickness wastes me.
Cattle in the shed; mead in the bowl.
The wise man desireth not discord.
Patience encompasseth understanding.

Cattle in the shed; mead in the bowl.
Slippery the paths; fierce the shower,
The ford bursts its banks.
The heart is deceitful above all things.
It is full of deceit, and wicked works.
There shall be grief when it is cleansed:
Exchanging for a little thing a great one.
Hell is in store for the base ones.
When the Lord judges on the long day
the false will be in gloom, the true in light.
Cups are held up; the attacker is ragged,
Men are merry over ale.
Withered the stalks; cattle in the shed.
I have heard a heavy-pounding wave,
loud between the beach and pebbles.
My heart rubs raw from depression tonight.
Oak-branch splitting; bitter the ash-taste.
Cowparsley is sweet; the wave laughs.
The cheek cannot hide the heart's grief.
Many a sigh tells against me
as follows my practice.
God allows no good for the hapless man.
Good to the hapless man may not be allowed,
only sorrow and care.
God will not undo that which He does.
Despite that which may be done in a prayer-house
a hapless man is he who reads it—
hated by man here; hated by God above.
The leper was a squire; he was a bold knight
in the court of a king.
May God be kind to the outcast. ➘

Author's note: I have tried to follow the source material closely while still striving to maintain a semblance of acceptable poesy in English. The author's gnomes are translated in seventeenth-century English, and, when the meaning of a line in the original Welsh is uncertain, I have consulted academic commentaries. For the interested reader, the standard editions are in Jenny Rowland's Early Welsh Saga Poetry *or Nicolas Jacob's* Early Welsh Gnomic and Nature Poetry, *though the latter offers no translation, only ample notes and a glossary.*

Lambing Season

I learned some of life's most important lessons from my father in the sheep barn.

NORANN VOLL

I WAS IN THIRD GRADE when Dad let me help him deliver lambs for the first time. I was not yet asleep that spring evening when I heard him in the hall filling up a bucket of water.

For the last year, I had been wishing to be part of this special event, and now he was saying the words I'd been waiting for: "Fill up another bucket, Nora, and make it hot. Mother, where are some soap and towels?"

That first night, the lambs came quickly. Dad was patient and alternately coaxed and pulled, talking gently to the ewe.

"You just work with the ewe; she knows you are here to help. Watch her eyes and her ears. They tell you just what's going on," he assured me.

Later: "Yes, of course they are in pain, but when you talk to them, they calm down and get to work. Remember this: it's a struggle to be born, and it's a

Spring lambs at Ribble Valley in Lancashire, England. Photograph by Lee Parkinson.

struggle to die. We all need help at the beginning of our lives, just like we need it at the end."

When the miracle of twins lay there in the sawdust and the ewe licked them off, urging them to stand for their first meal, I looked at Dad. His eyes shone as blue as chicory flowers and his face was wet with tears. "No human can create a beginning, Nora," he said, "and no human can create an end. It's all in God's hands."

Over the rest of my elementary school years, and all the way through high school, Dad and I delivered lambs side by side in the pen.

Toward the end of a lamb's delivery, Dad would always assist the ewe. But in the anxious hours prior to a birth, or when a wounded animal was healing, or when we were uncertain if the mother would care for her newborn, he advised, "Let nature take its course. It's the hardest thing for us humans to do, but it's the best. We always want to go poking around, but then we muck everything up. Mother Nature knows how to take care of things."

I could hardly bear to let "nature take its course" the time a neighbor's dog chased two pregnant ewes into the woods and ripped a third ewe from her ear to her shoulder. Dad was furious and calm. He headed off in pursuit of the missing ewes, while I helped our family doctor, always willing to patch up animals as well as people, tend the wounded one.

We sewed all afternoon, stitching bits of torn skin and wool back together, and trying to will the ragged edges of flesh to adhere, to heal. The doctor was confident that the ewe would recover: "With this antibiotic she'll be fine. Now, just let nature take its course."

But I was not confident in nature this time. The ewe lay motionless late into the evening, refusing drink or food. Meanwhile Dad returned, downcast. He had been unable to find the lost sheep.

It was Pentecost, and that evening our community sang fitting selections from Mendelssohn's *Saint Paul* oratorio. Afterward, Dad and I walked to the barn for a final check on the wounded sheep. I held Dad's hand and confessed that I'd tweaked some of the lyrics of a chorale with the line "and guide the wanderer kindly home," because I wanted God to hear.

"He does hear about sheep, doesn't he?"

"Of course, but what did you sing?"

"I sang, 'And guide the lost *sheep* kindly home. The hearts astray that union crave, and *sheep* in doubt confirm and save.' I sang it for the sheep that are out in the woods tonight."

Dad said nothing for a while, but squeezed my hand tightly as we walked on. When we got to the barn, the wounded ewe was up and drinking.

"We are all lost sheep, Nora," he told me on the way home, "and we are all God's sheep, all the time. Even when we don't think we belong to anyone, he is still our Father. There are times when we think we'd be happier out of the pasture, but he knows that we are truly happy when our boundaries are clear and we are safe. And there are other times when we think we are lost, but most especially then he is holding us all the time. All the time."

Norann Voll writes about discipleship, motherhood, and feeding people. She lives with her husband, Chris, at the Danthonia Bruderhof in rural Australia. They have three sons.

A photograph of the author's father hangs in the barn at the Fox Hill Bruderhof.

"Does he guide sheep back?"

"That is God's job. Guiding back is his main work. He does it all the time."

The next morning, the two lost ewes stood bleating for food at the barn door.

In 1998, two weeks before my wedding, I delivered lambs with Dad one last time.

I knelt there in the sawdust with the feeling that this would be the last and most precious time we would welcome life into the world together. The ewe had a rough go of it. Dad suspected extra-large twins, and we worked late into the night with the exhausted ewe.

Finally, one large, perfect lamb was born.

We couldn't help laughing. The ewe turned; it seemed as if she was smiling at the funny humans. I looked at my father and thought again how much I loved him, and how much I loved this place where he had taught me what he knew about how life works. I looked at the bleached blond and auburn lock of hair which always fell over his forehead when he worked, at his tired but satisfied eyes, sapphire blue and glowing with the joy of the new life lying at his knees.

And in the dim barn light, under the swallow's nest, I noticed his face was wet with tears. ⌇

Spring lambs at Ribble Valley in Lancashire, England.
Photograph by Lee Parkinson.

Editors' Picks

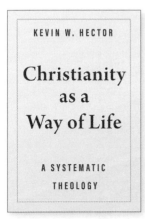

Christianity as a Way of Life
A Systematic Theology

By Kevin W. Hector
(Yale University Press, 308 pages)

Loving, singing, praying, lamenting, forgiving, being grateful, being generous, even laughing – these are just a few of the Christian practices that inform Kevin W. Hector's consideration of more traditional theological topics such as the problem of sin and evil, the saving mission of Jesus, and the end of the world. *Christianity as a Way of Life* shows that the discipline of systematic theology, which traditionally seeks to clarify Christian beliefs, can be reconceived as a study of the embodied, communal practices that make up the Christian life. From start to finish, the prose is clear and the uplifting vision of a life reoriented toward God shines through.

This book reunites doctrine and practice in a manner that many Christians will appreciate, but it also aims to persuade skeptics that theology has a legitimate place in the academy. It does this by arguing that theology offers understanding and wisdom that even scholars who are not devoted to Christianity can recognize as good. By interpreting Christianity not as a collection of beliefs but as a way of life among other ways to live in the world, theology has the potential to broaden anyone's understanding of life, regardless of his or her religious or nonreligious commitments. Moreover, by explaining how Christian practices can make people better by directing them toward recognizably good ends, theology may help even those who do not adhere to Christianity find some practical wisdom within it.

Yet such gains in broader legibility come with tradeoffs. Some Christian readers may find the distinctiveness of their faith underemphasized. Hector's book, which aims for academic legitimacy, might be helpfully complemented by other works that more boldly affirm aspects of Christianity that make it seem strange or even "foolish" (1 Cor. 1:27).

Hector's treatment of prayer provides a useful example. Hector describes prayer as an intentional practice of asking God for goods we desire, which can help us assess whether those goods align with our highest values. Over time, such a practice has the potential to transform our desires in beneficial ways. It helps us avoid taking good things in our lives for granted and heightens our gratitude. If we pray for others, this can increase our ability to love them, while curbing our egoism. In all of these respects, Hector makes a strong case for the practical wisdom of prayer. However, he does not give detailed discussions of what difference it might make to pray in the name of Jesus, what role the Holy Spirit plays in prayer, or the possibility of God speaking to us during prayer.

Though packaged for academic recognition, the Christianity presented here remains powerful, inspiring, and authentic. Informed by diverse traditions and brimming with fresh insights, Hector's book is a gift to theologians, Christians, and seekers of all kinds.

—Andrew Prevot, Georgetown University

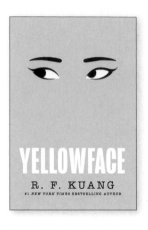

Yellowface
A Novel

*By R. F. Kuang
(William Morrow,
336 pages)*

The cover of *Yellowface* was no mistake. Against the plain mustard background, two human-sized eyes face the reader, foreshadowing the content.

June Hayward, plain and white, is casual friends with the beautiful and Asian Athena Liu. They are both authors – though Athena has seen bright success and June has seen none.

One night over drinks, Athena dies suddenly in front of June, leaving a first draft of her latest manuscript sitting on her desk. June calls 911, but not before stuffing the manuscript in her bag. Over the next weeks and months, she completes the manuscript, tightening, shaping, and omitting details from the novel about Chinese laborers during World War I. The book gets published and released to wide acclaim, while June struggles to come to terms with her theft, appropriation, plagiarism, and jealousy.

Author R. F. Kuang has said she based Athena on herself, and it's not difficult to see the similarities. They share nearly the same pedigree. Kuang is twenty-seven years old, with degrees from Oxford and Cambridge, and one in process at Yale. She's been nominated for the Hugo, Nebula, Locus, and World Fantasy awards and is a two-time *New York Times* bestseller.

One gets the sense that Kuang is laughing at herself and her industry, writing from June's perspective and asking the question, "Who gets to tell a story?" In one scene, June follows Athena around a museum where Athena picks at the bones of the stories of Asian lives during the war like a vulture, saving the tastiest morsels for her manuscript. In another scene, Athena co-opts June's traumatic event for her own short story. After Athena's death, June turns the tables, taking Athena's story as her own despite being white herself.

Yellowface is a story about the harms we do when harm has been done to us, about who gets to tell the stories of others, about what it means to succeed and to whom our success belongs, and about racism in publishing and its flip side, tokenism.

One cannot help but think of Narcissus, who died staring at his reflection in a pool of water. How we can avoid narcissism unless we take a

---

There are no heroes in *Yellowface*; everyone is complicit in the deception.

---

good, honest look at ourselves? How can we avoid the obsession with self that social media and success so easily produce?

Is it ironic that the eyes on the cover are what Athena calls "almond-shaped"? It is far too tempting to see my white self in them, like a mirror showing me all the ways I am complicit in an industry that too often centers and celebrates the wrong people.

There are no heroes in *Yellowface*; everyone is complicit in the deception: writers, editors, publishers, publicists, and readers. Yet, as one character says, "Writing is the closest thing we have to real magic. Writing is creating something out of nothing, it is opening doors to other lands. Writing gives you power to shape your own world when the real one hurts too much."

And who doesn't want that?

—*Lore Ferguson Wilbert, author,*
The Understory *(May 2024)*

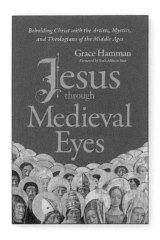

Jesus through Medieval Eyes

Beholding Christ with the Artists, Mystics, and Theologians of the Middle Ages

By Grace Hamman (Zondervan, 208 pages)

The year our first child was old enough to start understanding the words of Christmas carols, I wondered whether he realized that all the songs were about the same person. We sang about the Holy Infant, the Christ Child, the Messiah, Godhead, Incarnate Deity, Immanuel, the Baby in the Hay, the King of Angels. God became man: it is a wonder so great that Christians have been trying to find words to describe it ever since it happened. In *Jesus through Medieval Eyes*, Grace Hamman suggests that learning to look with the perspectives of Christians who lived long before us might enlarge our understanding of this mystery.

The book examines seven depictions of Christ that emerged in medieval Europe. Some of these images of Christ are still familiar today, but it's refreshing to encounter them in the vivid words and pictures of medieval writers and artists, whether Christ appears as the Knight who accompanies us in the costly acts of love we are called to, the Judge who is both almighty and merciful, the Mother who "gathers her chicks" as described in Matthew 23, or the Bridegroom from the Song of Songs.

Hamman uses her own experiences and outlook to guide the reader into the foreign country of the past. The result is a double refraction: some of her reactions to the source material were unexpected to me, although not so unexpected as the more startling fruits of the medieval imagination. Still, the earnestness and piety of the medieval Christian, especially in contrast with the pervasive irony and detachment of contemporary culture, is very touching. It was an age, Hamman writes, "saturated in beauty and love of Christ." She urges readers to encounter the ideas of our medieval brothers and sisters humbly and with childlike openness after the example of Julian of Norwich "reverently beholding" the mysteries of our Lord.

The book itself is attractively designed. The cover featuring gilded images of saints and the section of vibrantly reproduced artwork add a valuable dimension to the book, since many of the

---

Hamman uses her own experiences and outlook to guide the reader into the foreign country of the past.

---

themes discussed are highly visual. Each chapter concludes with a prayer in the words of a medieval Christian that brings the material into focus.

The book encourages a renewed sense of awe toward Jesus. Just as we can be surprised seeing a beloved person in a new setting and realize something new about him or her, so we might glimpse Jesus better by seeing him through the eyes of the people who built cathedrals (but who also burned heretics; not all ideas deserve to be preserved). Toward the end of the book, Hamman writes, "Like catching a piece of reflection in a broken mirror, each representation catches and renders an aspect of a Christ bigger, more beautiful, more glorious than any of them could separately communicate."

This book contains many astonishing and lovely facets of the Name that is beyond all names, who is also, in the words of a twelfth-century carol, "Jesus, our Brother, strong and good."

—*Marianne Wright,*
Seasons of Community Living *Substack* ❧

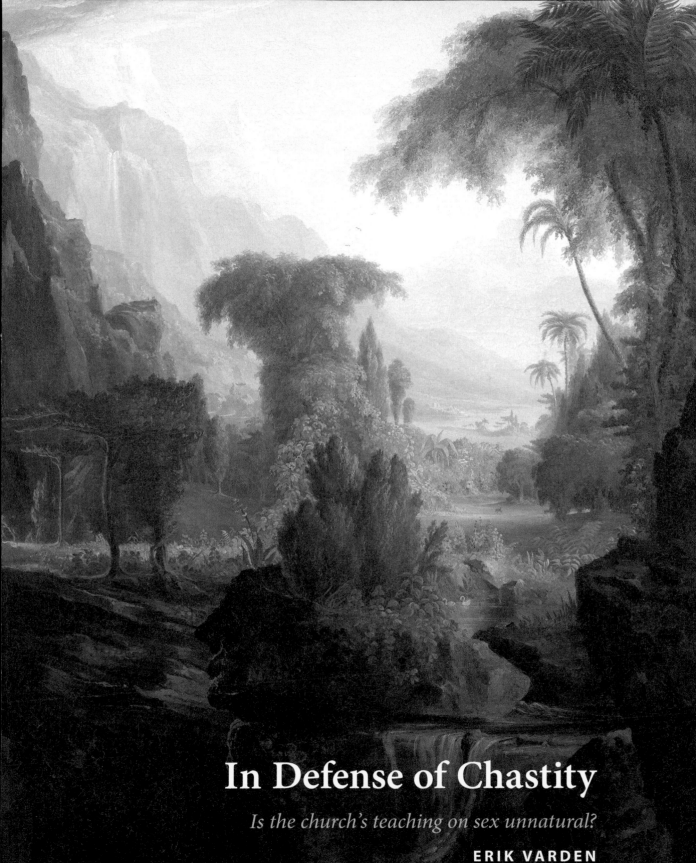

In Defense of Chastity

Is the church's teaching on sex unnatural?

ERIK VARDEN

GUNNEL VALLQUIST, Marcel Proust's first Swedish translator, once inscribed for me her diary of the Second Vatican Council, which she covered from start to finish as a correspondent for the Swedish press. She wrote out Galatians 5:1: "For freedom Christ has set us free." That phrase represented for her, as it must for any Christian, the core of the gospel and thereby of the church's mission.

Does the church's teaching on sex and chastity liberate? Many people think it does not. In taking a closer look at the question, however, a couple of distinctions are called for.

We as a society are muddled as to what it means to be liberated: to be "free." We ordinarily think of freedom as scope to do what we feel like. We think in terms of freedom *from*, not of freedom *unto*. In Christian terms, freedom is about enabling commitment. The biblical view of human nature, evidenced in Christ, regards the human being as essentially relational, defined by self-giving and commitment. On this account, unhindered pursuit of momentary inclinations is not freedom. It is enslavement to whim, which, empirically speaking, rarely produces lasting happiness. Sensational thrills can come of it, true, but they are not much of a foundation on which to construct a life.

Secondly, the freedom Paul speaks of is of a specific sort: that for which Christ has set us free. This freedom presupposes a call to self-transcendence. The finality of life, biblically speaking, is not limited to present thriving. Such thriving is a good, but a radically incomplete one.

Syriac Christian writings talk about the "robe of glory" – the God-given clothing that Adam and Eve put aside for "garments of skin" when they left Eden for a fallen, broken world. When we posit our present thriving as the be-all and end-all of existence, we sew ourselves up in those "garments of skin," locking ourselves into the limited, fallen self-understanding we assumed after the Fall. We lose sight of the "robe of glory" that alone reveals the sense of our desires and alone bears the promise of satisfying them.

Holiness, life everlasting, the resurrection of the body: these notions do not feature much, now, in people's thinking about relationships and sexuality. We have become alienated from the mindset that brought about the soaring verticality of the twelfth century's cathedrals, houses holding the whole of life while elevating it.

Was not a proposal recently made to fit a swimming pool on the rebuilt roof of Notre-Dame of Paris? It seemed to me apt. It would symbolically have re-established the dome of water that sealed earth off from heaven on the first day of creation, before God's image was manifest in it (cf. Gen. 1.7).

Whatever fragment of mystery might remain within the church itself would have been performed beneath the splashing of bodies striving to perfect their form. The parable would have been significant.

Once the supernatural has gone from Christianity, what remains? Well-meaning sentiment and a set of commandments found to be crushing, the profound character of the change they were meant to serve having been summarily dismissed.

Erik Varden is bishop of Trondheim and former abbot of the Trappist monastery of Mount Saint Bernard in Leicestershire.

It is time to effect a *sursum corda*, to "lift up our hearts," to correct an inward-looking, horizontalizing trend in order to recover the transcendental dimension of embodied intimacy, part and parcel of the universal call to holiness. Of course we should reach out to and engage those estranged

church went on to make sense of what it means to be a human being and to show how a humane social order might come about.

Today, this sense of who Jesus Christ is has been eclipsed. We still affirm that "God became man." But we largely apply the dictum in reverse,

> The church, surely, is called to provide the compass by which people of good will might orient themselves in times of confusion, not to run after the crowds like a puffing old spaniel striving to keep up with the hunt.

by Christian teaching, those who feel ostracized or consider they are being held to an impossible standard. At the same time we cannot forget that this situation is far from new.

In the early centuries of the church, there was colossal strain between worldly and Christian moral values, not least concerning chastity. This was so not because Christians were better – most of us, now as then, live mediocre lives – but because they had a different sense of what life is about. Those were the centuries of subtle controversies over what the incarnation really meant. Relentlessly, the church fought to articulate who Jesus Christ is: "God from God" yet "born of the Virgin Mary"; fully human, fully divine. On this basis the

projecting an image of "God" that issues from our garment-of-skin sense of what man is. The result is a caricature. The divine is reduced to our measure.

The fact that many contemporaries reject this counterfeit "God" is in many ways an indication of their good sense. What a contrast with earlier times. Nicholas Cabasilas, who lived at the time of the great medieval mystics Walter Hilton and Julian of Norwich, writes: "It was for the New Man that human nature was originally created; it was for him that intellect and desire were prepared. We received rationality that we might know Christ, desire that we might run towards him. For the old Adam is not a model for the new, but the new a model for the old."

In the midst of present perplexity, with the church weighed down by a history of abuse, with society's deconstruction of categories that, just yesterday, we thought normative, and with no shortage of people who, like the peers of Isaiah, "put darkness for light, sweet for bitter" (Isa. 5:20), we need to be recalled to this perspective.

The Desert Fathers' shortest dictum tells us to "look up, not down." The advice is sound. The church, surely, is called to provide the compass by

which people of good will might orient themselves in times of confusion, not to run after the crowds like a puffing old spaniel striving to keep up with the hunt.

This is not to say that the church should condemn the world. "Neither do I condemn you." These words of Jesus addressed to the woman caught in adultery remain a norm to which any ambassador of Christ is held. So do the words that follow: "Go your way, and from now on do not sin again" (John 8:11). Avoid wrong turns! The freedom for which Christ has set us free is freedom to follow him and to obtain the blessedness he has in store for us, not to get lost in the woods.

The Christian proposition of chastity is unattractive when put forward with rage, an attitude betraying self-righteousness on the part of its proponents. God, meanwhile, as John 8 shows, regards the affairs of human hearts and

bodies with illusionless patience – though let us remember that "patience" is more than a capacity to hang on and wait; at its heart is the root *patior,* which means "I suffer."

Christ does not flee from our contradictions. He does not shun in disgust the world of lusts and instant hopes that Siddhartha, in Herman Hesse's novel of the same name, calls the world of "people-who-are-like-children" (I once heard a seasoned confessor say, "You know, there are no

> **The Christian condition is the art of striving to answer a call to perfection while plumbing the depth of our imperfection – without despairing and without giving up on the ideal.**

adults, only children"). He enters that world and calls out to us, "Adam, where are you?" Sometimes he calls just by looking at us all-knowing, grieved at our estrangement, but not despairing of us.

We easily forget that God has hope for us. He knows we need to grow, and to grow up. The Church Father Irenaeus of Lyon presents Adam and Eve as children in the garden: "Man was a child, not yet having his understanding perfected; wherefore he was easily led astray." But therefore, too, he had potential freely to grow, learn, and change.

A Christian view of human nature is dynamic. Yes, of course we are conditioned by factors not subject to our choice; of course we carry gifts and wounds of all sorts; these condition us, but do not determine us. What determines a life is not the place from which it begins but the goal toward which it moves. If we believe, like Cabasilas, that the new Adam is a model for the old, we shall live prospectively, drawn by God's patient hope for us.

We are in the same boat. The second-century apologetic *Epistle to Diognetus* stresses that, left to ourselves, we are all borne along by "disordered impulses [ἀτάκτοις φοραῖς], carried away by desires and cravings." For some, disorder will be more patently "objective" than for others. But all of us are called to re-orient our lives toward the ultimate end revealed in Christ. In him fullness of life awaits us. Nowhere else.

After the death of the theologian Cardinal Jean Daniélou in 1974, Gunnel Vallquist wrote an essay on "The Daniélou Mystery." She had known the cardinal very well. She set out, though, not from private recollection but from a subject of public lechery: How had it come about that he, a prince of the church, had died of a stroke in the flat of a Paris prostitute, his wallet stacked with cash? Vallquist points out that Daniélou had long ministered to women who worked the streets of Batignolles. He helped them with alms to care for their often-complicated networks of dependents. He carried on this work after being created a cardinal in 1969. For Daniélou, such contact, illumined by Christian friendship, with people considered to be beyond the pale, was no big deal. So sharp, writes Vallquist, was his awareness of the chasm that separates us all from the uncreated glory of God that the calculation of degrees seemed to him absurd. He did not, by being a friend to prostitutes, relativize Catholic teaching: he wrote strongly in defense of chastity. But he was not shy to visit and assist those who, to reach this ideal, had a long way to go. He was just acting like his Master.

The Christian condition is the art of striving to answer a call to perfection while plumbing the depth of our imperfection – without despairing and without giving up on the ideal. Cancelling the ideal is tantamount to turning cathedrals into swimming pools, to replacing Christ's personal call, "Come, follow me" (Mark 1:17) with a pre-printed message to "take your ease, eat, drink, and be merry" (Luke 12:19). The goal to which we are called lies ever ahead. To stagnate is deadly.

But what if I have no strength to walk? Well, I must learn to let myself be carried. Israel's exodus, that exemplarily tortuous voyage during which the people tried every trespass, issued in the profession: "Underneath are the everlasting arms" (Deut. 33:27). Providence, Israel saw, had carried them through thick and thin. Their realization corresponded to the oracle God gave when they stood on the Promised Land's threshold: "The Lord your God bore you, as a man bears his son, in all the way that you went until you came to this place" (Deut. 1:31).

The primary form of *ascesis*, self-discipline, required of a Christian is trust. By trust we give up illusory claims to omniscience. We give ourselves into God's hands and choose to be reformed according to his purpose. Only he can realize his likeness in us, uniting in a chaste whole the disparate factors that make up our history and personality.

An error Christians have often made is to assume that chastity is somehow normal, but no, it is exceptional. Virtue does not come easily to us: when we try to practice it, we find that sin's wounds cut deep. They condition us to fail of our purpose. Even as we labor to learn charity, patience, courage, and so forth, we must labor to become chaste, letting grace do its slow, transformative work. Short of dazzling exceptions, growth in grace, like other growth, is organic. It happens slowly, secretly, we know not how (cf. Mark 4:27). But it does, in time, bear fruit.

Athanasius, in *On the Incarnation*, marvels at Christians who practice celibacy. For him, their witness is a sign of the end times. Subsequently continence came to be taken for granted. Youths embracing the clerical or consecrated life were simply expected to be chaste, without always understanding what their physical passion, a gift from God, represents or how it might be channeled responsibly. Many have lived lives marked by division, as if the senses were pursuing an unruly life of their own to be either suppressed by force of will or anesthetized.

Marguerite Yourcenar observes with regard to her depiction of Mary Magdalene, in the short story collection *Fires*, that the process by which vulnerability, desire, and love are changed into supernatural attachment is not one of "sublimation" but of orientation. I concur with her that "sublimation" is "in itself a very unfortunate term and one that insults the body." What is at stake is something else: "a dark perception that love for a particular person, so poignant, is often only a beautiful fleeting accident, less real in a way than the predispositions and choices that preceded it and that will outlive it." How do we handle these?

Physical and affective impulses are ordered according to an attraction of soul made conscious through application of the mind. The integral reconciliation of our being ("integrity" was long a synonym for "chastity") presupposes a certain kind of motivating energy – an élan. In the Vulgate version of one psalm, the goal toward which Israel journeyed through the desert is described as "the desirable land" (Ps. 105:24, Vulgate). The typology is timeless.

The single ascetic counsel Saint Benedict gives about chastity is "*Castitatem amare,*" "Love chastity." Only what I love will change me beautifully. Behaviors prompted by fear or disdain tend to disfigure. Love must be honed. The counsel on chastity is complemented by "*Ieiunium amare,*" "Love fasting." To refrain from feeding an appetite, even a physical hunger, can be a way of learning to love in an ordered, fruitful way.

I stress this aspect of learning. The Law, wrote Saint Paul to the Galatians, is, in the Douai-Rheims translation, a "pedagogue" (Gal. 3:24). The epistle's rhetoric makes us view the term critically. But to have a reliable pedagogue is splendid. In virtue, as in science and wisdom, we need to be taught. Our conscience must be formed. The point of Christian moral teaching is to outline a process

of learning conceived of in terms of conversion and *ascesis*, a term stemming from the Greek word for "exercise": a fitting metaphor in our society. The goal is freedom and thriving. To learn in this way is to see myself in terms of a reality that exceeds me. It is to be freed from imprisonment in my own limited notions.

Grounded in truth, we can reach immense stature. The life that pulsates in us carries an echo of God even when bogged down in self-destructive patterns. Saint John Climacus, abbot of Mount Sinai during the reign of Pope Gregory the Great, speaks of the maturing he has observed in people caught up in what nowadays we would call sexual addiction:

> I have watched impure souls mad for physical love but turning what they know of such love into a reason for penance and transferring that same capacity for love to the Lord. I have watched them master fear so as to drive themselves unsparingly towards the love of God. That is why, when talking of that chaste harlot, the Lord does not say, "because she feared," but rather, "because she loved much," she was able to drive out love with love (ἔρωτι ἔρωτα διακρούσασθαι).

It is an astonishing statement. It effectively dismantles the view that would separate spiritual *eros* – love – from carnal *eros*. For Climacus, they belong to a single continuum. He calls on "the chaste harlot" to witness his thesis. This perspective does not "insult the body." It offers neither sublimation nor appeasement. It acknowledges a flicker of eternity in passion. Even disordered *eros* can kindle a sanctifying love of God that drives out fear. Nothing is beyond God's ordering power. Nothing in man is unredeemed. Everything natural to man is made in view of the robe of glory. The new Adam waits to embrace the old. The garments of skin are lent to us for a while, to warm and protect us. Then we are to leave them behind. ➤

Excerpted from Erik Varden, *Chastity: Reconciliation of the Senses* (Bloomsbury, 2024). Reprinted by permission.

Squall

A front of thunderstorms had sought you out.
It vowed to run a diabolical
black line through all that you were sure about—
the ordinary, sane, the sensible.
You raced to get the loose stuff off the lawn,
with purpose rearranged and stacked the chairs,
relieved, almost, when the phenomenon
of gray-green storm clouds simplified your cares.
And though it couldn't miss, it kind of did.
Darkness at noon gave way to sun at one.
Catastrophe and doom had been short-lived.
Embarrassed that your fears were overblown,
you faced your mundane day-to-day concerns,
vaguely upset that normalcy returns.

ROBERT W. CRAWFORD

Elicia Edijanto, *Clouds*, graphite on paper, 2023.

Earthworks

*At a community garden in Detroit, kids help
grow food to feed the hungry.*

CASEY KLECZEK

DRIVING THROUGH Detroit's lower east side you see derelict shops, rotting Victorian homes, dilapidated factories slated for demolition, ragweed and lovegrass growing violently over sidewalks, and creeper vines enveloping telephone poles and old street lights, reclaiming the land. You also see dazzling street art on the crumbling wall of an old school or church, admonishing the doubtful that "Detroit never left," designers restoring the Ford family's historic homes to their old glory, Michelin-star restaurants donating to local charities, and, among all these sidewalk-crack dandelions, Earthworks Urban Farm.

Turning onto Meldrum at Mt. Elliott Cemetery, where some of the most elite Detroiters of the

Earthworks Youth Program students learn about different types of soil.

1800s were laid to rest, you pass a boarded-up Baptist church and two fire-singed homes where several temporary settlers are taking cover from the sun, lying on old pizza boxes. Then, there's Earthworks. Like a lantern, a statue of Saint Francis of Assisi greets you outside a hoop house as if to say, "This is the place."

Earthworks Urban Farm is a nearly two-acre certified-organic farm spread out over several blocks. The primary growing space, a half-acre behind a community food bank, boasts rows of radishes, arugula, mustard greens, potatoes, garlic, and spinach. There are also a greenhouse and hoop house for year-round production. Earthworks has orchards with cherry, apple, and peach trees, an area where they make their own compost, and an apiary hosting forty hives. Every year they send approximately four tons of food directly to the Capuchin Soup Kitchen, just a stone's throw away.

It was Brother Rick Samyn who dreamed these two wild, overgrown acres into a farm in the 1990s. He was a member of the order of Capuchin Franciscans, who first made their home in Detroit's Islandview neighborhood back in the 1880s, when the area was still farmland. They built St. Bonaventure monastery and traveled by foot, horse, and buggy to the far reaches of metro Detroit to offer confession and spiritual advice. During the Great Depression they expanded their focus when the neighborhood's poor started knocking on the door to ask for bread. "They are hungry; get them some soup and sandwiches," the doorkeeper, Fr. Solanus Casey (now beatified and on his way to canonization in the Catholic Church) was known for saying.

Their ad hoc soup kitchen grew as word spread; in time the lines grew to more than 2,000 people a day. This inspired the Islandview Capuchins to evolve through the years to respond to the needs of their neighborhood in whatever hunger arose.

When the neighboring Packard Automotive Plant, which had employed 40,000 at its peak, closed in the 1950s and thousands of Islandview residents were unemployed, they knew where to turn. When the KKK rose in prominence in the 1960s, burning crosses in neighborhood yards, the Capuchins delivered fiery homilies against racism and marched for civil rights. Through UAW strikes, race riots, deindustrialization, and the ensuing unemployment and poverty, the Capuchin brothers were a lighthouse.

So, in the 1990s, when the leader of the Capuchins' youth outreach programs, Brother Rick Samyn, was making a grocery list and a neighborhood child asked him, "What gas station do you get your groceries from?" it was a clear alert to neighborhood need. The Capuchins would do what they always did: they would feed the hungry.

"There was this food desert, like so many urban areas. No grocery stores. Not much in terms of gardens," explains Brother Gary Wegner, the current executive director of the Capuchin Soup Kitchen. "And so it started with Brother Rick wanting to give kids an opportunity to see where food really comes from."

At that time, nineteen of Detroit's neighborhoods were labeled food deserts by the Michigan Department of Agriculture. Over thirty thousand residents didn't have access to a full-line grocery store and 50 percent of households were food insecure, relying on corner stores, liquor stores, or fast-food chains to eat. People would have to travel miles away from their homes for adequate or healthy food, which posed a problem for the third of Motor City residents who didn't have access to a vehicle.

"After the riots in the sixties, people were moving out of the city in droves, so over time these vacant buildings were demolished and the lots left vacant," says Wendy Casey, director of Earthworks.

Casey Kleczek is a writer, documentary filmmaker, and mother based in Metro Detroit, Michigan.

"Deindustrialization, automation, industry consolidation, and disinvestment really hit the neighborhood particularly hard," explains Tim Hinkle, director of public relations for the Capuchins. "Schools closed. Shops closed. Grocery stores closed, including the one that was right on the spot where Earthworks is now." The population of Detroit dropped from two million in 1950 to 680,000 today. At one point, 37 percent of Detroit was vacant land. As residents left the neighborhood for the suburbs, supporting businesses closed, including the little grocery store on the plot of land across from the Capuchins. Eventually that store, along with most of the surrounding homes and businesses, was demolished. Aerial photographs from that time show urban core receding to urban prairie, a reality that yielded unlikely juxtapositions. Pheasants moved in. Deer are still frequent grazers at the Earthworks garden, and are becoming the latest nuisance in Detroit backyards. "It's an industrial city that has aspects of rural now," says Brother Gary.

"It was during a time when there was a lot of attention on Detroit and how to repurpose all of the vacant land," remembers Wendy. "What do we do with these urban centers that have been decimated by disinvestment?" Many real estate developers saw the situation as an investment opportunity, buying up properties in the thousands with the expectation that they had found the next great urban market. Instead, these properties continued to fall into disrepair, with neighbors helpless to beautify their own neighborhoods and improve their property values. But while many developers were strategizing about how to revive urban life, Brother Rick Samyn looked out at the two acres of industrial brownfield surrounding the monastery and he didn't see empty space. He saw a farm.

Detroit's Farming History

Detroit has a storied history when it comes to urban farming. During another economic crisis in 1893, Detroit's mayor Hazen S. Pingree became a major proponent of vacant land cultivation as a means of helping unemployed workers in the city, largely Polish and German immigrants fresh from the agricultural economy of Europe. As a railroad and dockworker strike embroiled the city and cries of "bread or blood" echoed outside his office, he came up with a way to provide "bread" for those most deeply affected by the economic crisis. He called it the "potato patch plan."

The plan was to let Detroit's poor residents garden on vacant land to grow their own food. There were plenty of skeptics, and editorial cartoons ridiculed the idea. But one year later, the critics were sheepishly silent. In its first year, nearly a thousand families raised $14,000 from their crops on 430 acres of formerly vacant land – potatoes, yes, but also beans, squash, pumpkins, string beans, cabbage, cucumbers, corn, and beets. Within four years, the program had over 1500 families participating, and was adopted in other cities: New York, Boston, Chicago, Minneapolis, Seattle, Duluth, and Denver. Pingree was invited to speak all over the country. At a talk he gave in Terre Haute, Indiana he said, "Until such a time where society has learned to 'do justice to all,' we must depend on the methods nearest at hand."

Earthworks Grows

The "methods nearest at hand" for Brother Rick were the land and any willing participants. He used the land across from the monastery, where a community food bank operated out of a warehouse on a corner of one of the acres. Its owners charged him nothing. He built a few raised beds and planted some staples – tomatoes, lettuce, cucumbers.

"In the beginning it was just going to be a community garden," explains Wendy. "Then he incorporated youth and the program just grew. It became two youth programs, one for younger and one for older kids, and then we added an adult training program and a market."

A few raised beds eventually became the 1.25-acre farm it is today. The team of one friar became a staff of five and a roster of over a hundred volunteers. All the farm's produce is donated to the Capuchin Soup Kitchen, which serves approximately 150,000 meals a year to Detroit residents experiencing hunger and often homelessness or unstable housing.

Today, volunteers come from all over the world. They've had several groups from France and Germany who were inspired by the work. "For better or worse, true or not, Detroit has become a symbol worldwide of urban decline in the United States," explains Brother Gary. "And so when you see things like Earthworks begin to flourish, that captivates the imagination and then they want to come and see it themselves."

Earthworks has expanded to include a farm training program, Earthworks Agriculture Training (EAT), begun in 2010 to provide Detroit residents with skills needed to succeed in agriculture. Around ten people take the course each year, receiving training on the fundamentals of growing food. Since its inception, several students have already created their own businesses using skills gained from the program, from catering to starting their own farms. One woman started a line of natural herbal skin care products. Another became a fruit and flower producer. A third started a class on flower arrangements and terrarium design. Most of the EAT participants come from the surrounding neighborhoods, including some of Detroit's most impoverished areas.

A Space to Escape

One of Brother Rick's earliest green-thumb disciples was a six-year-old boy with unrivaled energy, a quirky personality, and a closet full of superhero T-shirts – Tyler Chatman.

"I've been here for a *good* minute," says Tyler, now an adult. "I started off in their youth program and I just kept coming back because it was so interesting to me."

Seeing grown men in long brown robes who chose poverty and shuffled off for lauds and matins when bells rang was never a barrier to Tyler. "Other people in my neighborhood might have been curious why there were people walking around in robes, but I grew up going to church, so I knew what they were about."

Besides, Brother Rick had the knowledge Tyler was drawn by: a boundless capacity for all things gardening. "It was more than just showing us how to plant a seed and learn where food comes from. He took us around Detroit, to find more

Beaumont Hospital medical residents pull weeds and help pick produce.

soup kitchen and learn how to grow food and things like that, I could have been out in the streets doing something I'm not supposed to be doing and probably getting myself into some kind of trouble. Being outside in nature, putting my hands in the ground, doing some hard work, pretty much built the character that I am now."

That was twenty-one years ago. Today Tyler is the farm manager at Earthworks. He oversees the planting season, monitors pests and weeds, determines what crops need to be planted each year, manages the volunteers, and teaches the classes. Every day from 5 a.m. to 4 p.m., you can see Tyler bustling between the farm and the soup kitchen. "He even comes on his days off," chimes in Wendy, chuckling.

"I just love working with my hands. I like to be active," Tyler responds. "It's pretty incredible to be able to take a tiny little seed no bigger than the tip of my finger and it becomes this massive amount of fresh-grown produce. You can take a tiny tomato seed, put it into the ground, and you'll see a whole big stem of tomatoes growing from that one little seedling. You'll have twelve or thirteen tomatoes, and you can harvest those and it'll just keep coming back with more until it's time for its cycle to end." Tyler grins sheepishly, "It's a pretty marvelous job that I have here."

A Girly-Girl

Brittney Hughes was a self-professed "girly-girl" who considered dirt the stuff she had to clean off her car, and grass a nuisance to mow. That was before joining the EAT program. It wasn't exactly

areas with potential for different farms. We learned about pest control and how to maintain a garden through all stages of the growing season. He taught us things like how to take tomatoes and grind them up into tomato paste. We had a day where we all made our own fresh salsa from garden vegetables. There were days we would make pizza and pies. It was basically teaching us how to grow our food and how to prepare it – how to be self-sustainable."

But more than learning self-sustainability, Tyler found the farm was a space to escape alternatives. "I kept coming back," explains Tyler, "It was interesting to me. It was something to do as opposed to sitting at home playing video games." At the time, Tyler says there weren't many kids his age in the neighborhood. So coming to a blossoming orchard with a lively monk, a gaggle of volunteers, and other interested students made for a great alternative. "It's kept me out of trouble. Being in Detroit, things could be kind of rough. There were probably moments where if I didn't come to the

Earthworks Agriculture Training students prepare a row for seed.

a match made in heaven. Though she had grown up just blocks from Earthworks, she had never heard of it.

"The EAT program was truly not part of my nature. I would never have thought to put my hands into some soil on a daily basis," laughs Brittney. The director of the EAT program at the time, Marilyn Barber, had been a member of Brittney's church. When Marilyn was looking for participants for her program, the pastor recommended Brittney. Brittney had been working eight to ten hours a day in the stuffy, hair-sprayed air of a salon, styling hair and doing makeup. She loved it, but from the first few hours she experienced between the rows of crops at Earthworks, kneeling in the dirt, she found something she loved even more.

"This program truly changed my direction in my career, in life, and in the way I see myself. I grew up so much because of this," Brittney reflects. "I learned how to sustain myself and how to teach others."

In the EAT program Brittney and the other students learned not only how to grow food, process it, and market it–they also learned business management. Because of this Brittney was inspired to start her own green-based business. She's created beauty products that she sells at several events each year. "I've traveled and met wonderful people from all over the world and it truly changed my life," she says. "I never got to step outside my environment in such a beautiful way."

Speramus Meliora

Detroit's motto has an interesting backstory. Inscribed on its official seal are the Latin words *Speramus meliora; resurget cineribus*, though few residents know what they mean.

The fledgling city was just 104 years old when a fire swept through in June 1805. Established by the French explorer Antoine de la Mothe Cadillac in 1701, it had survived the French and Indian War, the Revolutionary War, and all the trials of establishing a city when a local baker's tobacco ashes carried in the wind to the hay in his stable. Within minutes his barn was consumed in flames. The fire spread quickly to the neighboring homes and shops, too quickly for the efforts of the "bucket

---

Detroiters faced these bare plots of land with little more than the seeds in their pockets and a collective desire for survival.

---

brigade" that Detroiters formed, passing buckets of water from the Detroit River. Their efforts were in vain; the fire razed the city by late afternoon. Amazingly, all six hundred residents survived, but they now faced an unthinkable reality.

They had nothing left but empty lots. Father Gabriel Richard, a pastor of Ste. Anne de Détroite church, organized food shipments in the immediate aftermath. One of the local judges immediately went to work drawing up plans for the reconstruction of the city. A chief justice and architect drafted an ambitious new layout for the new streets of Detroit. When some residents wanted to pack up and move downriver to Monroe, Father Gabriel urged them to stay and rebuild with the words *Speramus meliora, resurget cineribus*–"We hope for better things; it shall arise from the ashes."

At crucial points throughout the two centuries since, Detroiters have faced these same bare plots of land with little more than the seeds in their pockets and a collective desire for survival and asked themselves the questions: eat or starve, stand or despair, surrender or endure. For two centuries, they've answered those questions with the motto: "We hope for better things; it will rise from the ashes." And for the last twenty-five of those years, Earthworks has been one of those "better things." ⤳

Watching the Geminids

Sky-watchers have a chance to think about time differently.

MAUREEN SWINGER

IT'S 3:00 A.M. on a winter night, and my fifteen-year-old daughter and I are lying on our backs on the picnic table in our yard. It's twenty degrees, but we have forgotten that we're supposed to feel cold.

We are doing this because tonight is the peak of the Geminids, known to be the most prolific meteor shower of the year. We're doing this because she is leaving home, and I'm not ready to part with her yet.

All right – she'll only be gone for five months. She's traveling to England as an exchange student; if I had that chance at fifteen, I'd want my mother to say, "Go! Go!" And I am – I *am* saying that. Cliffs of Dover, forests of bluebells, castles by the sea, streets Jane Austen walked. A boarding house under the eaves of a manor. Still, she won't be here. We won't have mother-daughter talks or walks or book chats or movie nights, and it doesn't help that all the experienced empty-nesters are saying,

Maureen Swinger is a senior editor at Plough. *She lives at the Fox Hill Bruderhof in Walden, New York, with her husband, Jason, and their three children.*

Dai Jianfang, *Geminid Meteor Shower,* 2021.

"Five months are nothing!" It's the beginning of the exodus. Don't talk me out of my tears.

We started watching stars when she was a few weeks old, hardly six pounds, and, according to her dad, "nothing but eyes." At age one, she would lie back against me in the Adirondack chair, way past bedtime on a June evening, and point first up at the still sky, and then over to the wood's edge where the fireflies were anything but still. She didn't make a sound; her eyes did all the talking. When she was two, and talking a mile a minute, she would still just gaze wordlessly upward at star-time; only once she said, quietly to herself, "God."

I took my son out in winter, because he had croup and cold air is much easier to breathe. We'd sit in the same place in the yard, wrapped in layers of quilts, and he'd stab his chubby finger upward between barking coughs to inform me, "'Tar! 'Tar!" This would go on for some time, since he felt the need to identify each one.

For a while we forgot about these late-night escapades; daytime life got busier, and certainly by the time a surprise third child came along five years later, we were glad enough for everyone to fall into bed at the appointed hour, including ourselves.

It's not as if life has now acquired a calm and stately pace. But as my eldest prepares for her first adventure without us, I think back often to my own childhood and the many gifts I was bestowed: my parents' example, my grandmother's quiet love, the strength of a believing community, and the lessons taught by one teacher in particular who showed us beauty in cloud patterns, bird migration, constellations. Almost all I know about the sky I learned from him.

Which is to say, I learned wonder. I don't have a scientific mind, despite having attended lectures about gravitational pull, the time-space continuum, black holes, and other mysteries of the universe, inasmuch as anyone can explain them.

Instead of explaining, I'd rather go out at night with whoever can pry their eyes open, just to look up. It's so very quiet in the predawn hours, and the stars are startlingly big and bright, as if they are closer to us than they were in the evening. Even one night in November when my eldest and I spent almost an hour waiting for some Leonids to make a faint appearance, we didn't miss what didn't come; the silence and the sense of the whole majestic universe turning above us made us both think, "God."

And now, on the night of December 13, do the Geminids show up? They do! With over seventy-five meteors an hour flinging out from asteroid 3200 Phaethon, and a clear night with no moon, we watch in awe as the slashes of light appear, anywhere at all in the sky, falling, flying, under-scoring constellations for split seconds, tangling in the black branches of the giant white oak.

We are quiet for a long time. Lines of a song come to mind – a verse from Marjorie Pickthall's poem "Stars":

And all the lonelier stars that have their place,
Calm lamps within the distant southern sky,
And planet-dust upon the edge of space,
Look down upon the fretful world, and I
Look up to outer vastness unafraid
And see the stars which sang when earth
  was made.

These are the lights that we look to see now, the ones that watched when the foundations of the earth were laid, "while the morning stars sang together and all the angels shouted for joy" (Job 38:7).

I like to think that star-music will always connect this daughter and me, without the need for words. She'll be watching the same night sky, only five hours earlier than me, for a little while. She'll be back in time for the Perseids.

We drift indoors, and she goes back to bed, while my son emerges long enough to see three unmissable shooting stars; ('Tar!) there, he's done it. But youngest daughter's eyes pop open; she bursts out the back door to spend the next half

hour pointing and shouting rapturously to me and the world at large: "There! Over there! Did you see it? It was so bright!" This gives way to reflections about where among the stars heaven is, and whether each star has an angel, and how fast they can travel between stars, and perhaps the shooting stars *are* angels.

Not content with peak night, she makes me promise to get her up again on the next night, and it is every bit as beautiful. I let her teachers know she might be tired and cranky, and she is. I'm sorry they have to pay for our nights of wonder. But not too sorry. After all, how often do these chances come around? While my first stargazing partner is away, I'm relieved and thankful to look upward with this eager child, toward "lonelier stars that have their place."

Sky-watchers have a chance to think about time differently. You might decide on a whim to go out and see what the stars were doing four thousand light years ago. You might also want to know that every thirty-three years those often disappointing November Leonids blaze in a perfect storm of hundreds – even thousands – of meteors an hour. The next one is in 2033 and my youngest daughter and I have it on our calendar, even if the rest of the family is out in orbit. Wherever they are, I'm sure they'll be looking up.

I trust my children to believe with Walt Whitman that "a leaf of grass is no less than the journey-work of the stars," that a picnic table in a freezing winter yard is also a floating board in a spinning galaxy, that we too belong to the one who made everything else moving through space. I trust them to travel out with his guidance, like 3200 Phaethon, and circle back again every so often.

For a visual calendar of meteor showers, see *meteorshowers.org*.

Rocky Raybell, *Geminid Meteor from Keller, Washington*, 2016.

PLOUGH BOOKLIST

New Releases

Tears of Gold
Portraits of Yazidi, Rohingya, and Nigerian Women
Hannah Rose Thomas

This debut art book by British artist and human rights activist Hannah Rose Thomas presents her stunning portrait paintings of Yazidi women who escaped ISIS slavery, Rohingya women who fled violence in Myanmar, and Nigerian women who survived Boko Haram captivity, alongside their own words, stories, and self-portraits. A final chapter features portraits and stories of Afghan, Ukrainian, Uyghur, and Palestinian women.

"I very much hope that this beautiful book will help enable these women's voices to be heard, as well as to highlight the issue of the persecution of religious and ethnic minorities in general."
  —HRH The Prince Charles, former Prince of Wales

Hardcover, 128 pages, ~~$49.95~~ **$34.96 with subscriber discount**

Come Again, Pelican
Don Freeman

From the creator of *Corduroy*, a newly restored classic picture book that celebrates a child's bond with the natural world. Every summer Ty's family came to camp in their trailer at the same beautiful spot on the white sand dunes by the ocean. And every year, as long as Ty could remember, the same old pelican had welcomed them. This year, as soon as the trailer was parked, Ty pulled on his shiny red wading boots and ran with his fishing pole to look for his friend.

Hardcover picture book, 44 pages, ~~$18.95~~ **$13.26 with subscriber discount**

Poets in Nature

Poems to See By
A Comic Artist Interprets Great Poetry
Julian Peters

This stunning anthology of favorite poems visually interpreted by comic artist Julian Peters breathes new life into some of the greatest English-language poets of the nineteenth and twentieth centuries. These are poems that can change the way we see the world, and encountering them in graphic form promises to change the way we read the poems.

Includes poems by Emily Dickinson, Langston Hughes, Carl Sandburg, Maya Angelou, Seamus Heaney, e. e. cummings, Robert Frost, Dylan Thomas, Christina Rossetti, William Wordsworth, William Ernest Henley, Robert Hayden, Edgar Allan Poe, W. H. Auden, Thomas Hardy, Percy Bysshe Shelley, John Philip Johnson, W. B. Yeats, Gerard Manley Hopkins, Edna St. Vincent Millay, Tess Gallagher, Ezra Pound, and Siegfried Sassoon.
Hardcover, 160 pages, ~~$26.00~~ **$18.20 with subscriber discount**

Water at the Roots
Poems and Insights of a Visionary Farmer
Philip Britts

In a society uprooted by two world wars, industrialization, and dehumanizing technology, a revolutionary farmer turns to poetry to reconnect his people to the land and one another. Amidst these great upheavals, his response – to root himself in faith, to dedicate himself to building community, to restore the land he farmed, and to use his gift with words to turn people from their madness – speaks forcefully into our time.
Softcover, 179 pages, ~~$16.00~~ **$11.20 with subscriber discount**

The Gospel in Gerard Manley Hopkins
Selections from His Poems, Letters, Journals, and Spiritual Writings
Gerard Manley Hopkins and Margaret R. Ellsberg

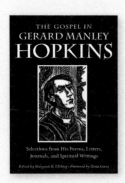

We love Hopkins not only for his literary genius but for the hard-won faith that finds expression in his verse. Who else has captured the thunderous voice of God and the grandeur of his creation on the written page as Hopkins has? Seamlessly weaving together selections from Hopkins's poems, letters, journals, and sermons, Ellsberg lets the poet tell the story of a life-long struggle with faith that gave birth to some of the best poetry of all time.
Softcover, 268 pages, ~~$19.95~~ **$13.97 with subscriber discount**

Resurrection Season

Bread and Wine
Readings for Lent and Easter

Dietrich Bonhoeffer, Dorothy Day, Søren Kierkegaard, C. S. Lewis, Eberhard Arnold, Fyodor Dostoyevsky, George MacDonald, Sadhu Sundar Singh, Thomas Merton, N. T. Wright, William Willimon, and others

Culled from the wealth of twenty centuries, these accessible selections are ecumenical in scope, and represent the best classic and contemporary Christian writers.

"Has there ever been a more hard-hitting, beautifully written, theologically inclusive anthology of writings for Lent and Easter? It's doubtful." —*Publishers Weekly*

Hardcover, 430 pages, ~~$26.00~~ **$18.20 with subscriber discount**

Easter Stories
Classic Tales for the Holy Season

C. S. Lewis, Elizabeth Goudge, Leo Tolstoy, Jane Tyson Clement, André Trocmé, Alan Paton, Oscar Wilde, Ruth Sawyer, Anton Chekhov, Selma Lagerlöf, Claire Huchet Bishop, and others

A treasury of read-aloud tales selected for their spiritual value and literary integrity.

"This thoughtfully curated collection is remarkable for its range and breadth. . . . Definitely read these stories at Easter, but keep the book close and pull it out whenever you and your family need a reminder of the great Easter themes of transformation, reconciliation, and the triumph of life over death." —*National Catholic Register*

Softcover, 383 pages, ~~$18.00~~ **$12.60 with subscriber discount**

The Scandal of Redemption
When God Liberates the Poor, Saves Sinners, and Heals Nations

Oscar Romero

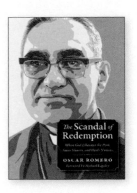

These selections from Archbishop Oscar Romero's diaries, letters, homilies, and radio talks invite each of us to align our own lives with the way of Jesus that lifts up the poor, welcomes the broken, wins over enemies, and transforms the history of entire nations.

"Romero does not speak from a distance. He does not hide his fears, his brokenness, his hesitations. It is as if he puts his arm around my shoulder and slowly walks with me. He shares my struggles. There is a warmth in his words that opens my heart to listen."
—Henri J. M. Nouwen

Softcover, 139 pages, ~~$12.00~~ **$8.40 with subscriber discount**

Maximus the Confessor

When a Byzantine monk looked at nature, he saw signs that point to something beyond.

SUSANNAH BLACK ROBERTS

IN AD 653 Byzantine Emperor Constans II ordered the arrest of Pope Martin I, and with him, an elderly monk named Maximus. The pope had been elected four years earlier against the emperor's will, and had called a council to deal with the most pressing theological issue of the day: Did Christ possess a human will as well as a divine one? The council said yes; the emperor disagreed. Put on trial, the pope and Maximus both refused to recant. Sent into exile, Martin died within a few years, but Maximus lived on, until 662, when he was brought to trial once again. Again, he refused to recant. He was tortured, his tongue cut out and his right hand cut off, so that he could no longer speak or write. Cast into the fortress of Schemarum, in what is now Georgia, he died soon afterward.

What had he written with that hand? What had he spoken with that tongue? Who was this man?

According to tradition, Maximus was born to a noble family in Constantinople and became first secretary to Emperor Heraclius, grandfather of the emperor who persecuted him. He left imperial service to become a monk, and eventually abbot, at a monastery in Chrysopolis.

His theological vision was rich and broad. Of particular interest here is his idea of nature as a book. As a Christian, Maximus was an heir to the traditions of the Hebrews, but he was also addressing the questions of Greek philosophers such as Thales and Aristotle, who had tried to explain the world through natural causes. Natural phenomena, Maximus agreed, were not divine in themselves, nor caused by gods, and could be investigated on their own terms. But why, he asked, are we able to understand ourselves and the natural world in the first place? Only because that world was made, said Maximus, by the God who created us in his image, capable of comprehending the universe or at least able to perceive its patterns – that is, able to read the book of nature.

For Maximus, just as for Thales and Aristotle, nature is emptied of its divinities. It is no longer a self-enclosed system of petty gods of springs and trees and ocean and earth and sky. But he also didn't view nature as a self-enclosed system of cause and effect, as Thales and his fellows (and their materialist heirs) would posit. Nature is not just something to which we can apply science. His reading of nature was not an attempt to stretch her on the rack and make her give up her secrets.

Rather, when natural phenomena cease to be gods, they become something better: *signs*. Nature is not self-enclosed at all. It is not self-explanatory, but it does explain. It points, very precisely, beyond itself, to the one who made it – who made all natural phenomena and any supernatural ones.

The apostle Paul says of the wise, "For what can be known about God is plain to them, because God has shown it to them. For his invisible

Susannah Black Roberts is a senior editor of Plough. *She and her husband, Alastair Roberts, split their time between New York City and the United Kingdom.*

attributes, namely, his eternal power and divine nature, have been clearly perceived, ever since the creation of the world, in the things that have been made" (Rom. 1:19–20). Expanding on this, Maximus says natural phenomena are signs pointing toward the truth behind nature: Christ the Logos, who "wills always and in all things to actualize the mystery of his embodiment" (*Ambigua* 7.22).

But that's not all. This God, who made man in his image and capable of reading his signs, does something even more shocking. He plunges fully

into nature, into the womb of a Jewish girl, taking on human nature – not by any half-measures, but fully. Fully God and fully man. And then he dies, and does not stay dead. By doing this, he makes possible the final step: he brings us humans up with him into the life of God. Eastern Orthodoxy calls this *theosis*, coming into union with God.

For Maximus, this possibility is shattering in its implications. Adam was given care over the earth: when he fell, the cosmos fell too, becoming "subject to decay" (Rom. 8:20–22). Christ reverses this. He is not only the "divine light" that illuminates the book, scriptural and natural; he also reveals our vocation in this world. Maximus reasons *from* the Incarnation *to* our vocation as human beings, which is to mediate and unite all the "divisions of nature" we experience, through rational understanding and loving action. So, for Maximus, our *theosis* is part of the Word's ongoing incarnation:

> Of these good things [in Christ], the modes of the virtues and the inner principles of what can be known by nature have been established as figures and foreshadowings, through which God always willingly becomes man in those who are worthy.

This could be taken in an unorthodox way, as suggesting that any of us could be Christ. Maximus, always orthodox, is careful to rule out such interpretations. Rather, in the New Adam, Jesus, humans are restored, drawn up into the life of God: and the whole cosmos, the whole of the natural world, is drawn up after them. It is natural for us to be supernatural, in other words. The human vocation is to extend the incarnation into the cosmos. We do this in every instance of cooperation with grace. Every act of perceiving the good, of understanding, of piety, of love continues Christ's incarnation in all things. Here Maximus is merely confirming scripture, which states plainly:

> Do not lie to one another, seeing that you have put off the old self with its practices and have put on the new self, which is being renewed in knowledge after the image of its creator. Here there is not Greek and Jew, circumcised and uncircumcised, barbarian, Scythian, slave, free; but Christ is all, and in all. (Col. 3:9–11)

Maximus is precise about these things, but bold. This, says Maximus, is *theosis*:

> the whole man pervading the whole God, and becoming everything that God is, without, however, identity in essence, and receiving the whole God instead of himself, and obtaining as a kind of prize for his ascent to God the absolutely unique God, who is the goal of the motion of things that are moved, and the firm and unmoved stability of things that are carried along to Him, and the limit (itself limitless and infinite) of every definition, order, and law, whether of mind, intellect, or nature. (*Ambigua* 41.5)

This, then, is the destiny of the cosmos – through us. This is what it is all pointing toward, what all those signs have been trying to tell us all along: the marriage of heaven and earth. It's as though we, the Bride, are pulling all of it, all the snails and woods, geese and galaxies, along with us as we walk up the aisle toward the Bridegroom. ➤

With thanks to Dr. Jordan Wood for consultation.

Constans II, *left,* having Maximus beaten, *right,* for his Christological views. Miniature from the twelfth-century Manasses Chronicle.